Ecosystem Conservation and Management

Marino Gatto · Renato Casagrandi

Ecosystem Conservation and Management

Models and Application

 Springer

Marino Gatto
Dipartimento di Elettronica
Informazione e Bioingegneria
Politecnico di Milano
Milan, Italy

Renato Casagrandi
Dipartimento di Elettronica
Informazione e Bioingegneria
Politecnico di Milano
Milan, Italy

ISBN 978-3-031-09482-8   ISBN 978-3-031-09480-4   (eBook)
https://doi.org/10.1007/978-3-031-09480-4

This Springer imprint is published by the registered company Springer Nature Switzerland AG
The registered company address is: Gewerbestrasse 11, 6330 Cham, Switzerland

Augescunt aliae gentes, aliae minuuntur,
inque brevi spatio mutantur saecla animantum
et quasi cursores vitai lampada tradunt.

Some populations increase, others decrease,
and in a short while the generations of living creatures are changed
and like runners relay the torch of life.

Lucretius, De Rerum Natura, 2, 77–79

Preface

The scale of human impact on our planet has been increasing decade after decade. Economic indicators of wealth (such as income per capita or poverty gap ratio) have been improving, though not homogeneously over the world. However, many targets of basic SDGs (Sustainable Development Goals)—like eradicating extreme poverty, ending all forms of malnutrition and of preventable child deaths, eliminating gender disparities in education, and achieving universal and equitable access to safe and affordable drinking water for all—are still far from being reached. Also, there is concern that the functioning of our Earth may be imperiled by the increasing stress due to human activities. In particular, there has been an enormous increase in the emission of greenhouse gases due to fossil fuel and biomass combustion and agricultural practices in the last decades. Air and water pollution, overexploitation of natural resources (think of marine fish stocks and tropical forests), massive land-use change together with Global Climate Change (GCC) may decrease the biodiversity of our planet and impair the functioning of ecosystems. We know this has important economic and social consequences because the services freely provided by ecosystems to humanity are enormously valuable.

The globalization of trade and travel which makes the world more and more connected has also driven the increasing risk of alien species introduction and their diffusion into countries and continents where they were not present in the past. These species often have a competitive advantage over native species, or carry new parasites and diseases, or act as new predators against which native species are defenseless. In addition, many of the pathogens hosted by nonhuman animal species can be transmitted to humans too and be very dangerous, as proved by the present COVID-19 pandemic. Devising paths toward the sustainability of our biosphere will thus be a mandatory task for humanity in the near future. To effectively cope with these challenging problems, it is fundamental to develop appropriate quantitative models that allow researchers and decision-makers to (i) depict scenarios of ecosystem dynamics in response to different conservation strategies and conservation policies, (ii) evaluate the probability of different scenarios, and (iii) find optimal strategies and policies according to different management goals, which account for both monetary and nonmonetary environmental valuations.

This book introduces the basic modeling notions to that end and describes the application of the illustrated quantitative tools. The mathematics employed is kept to a rather simple level, and the theory is always complemented by examples derived from real data. The book is divided into four parts.

Part I, after reviewing the drivers of biodiversity decline and its impacts, describes the processes that may lead a species to extinction. The Allee effect and genetic deterioration are modeled via suitable deterministic and stochastic approaches. Then, the general problem of extinction risk due to demographic and environmental stochasticity is described and the pertaining models illustrated. The synergism of different drivers can then be accounted for by means of Population Viability Analysis (PVA). The main modeling tools of this part are birth-death processes and stochastic discrete-time equations.

Part II introduces space as a basic dimension for the study of alien species invasion and the analysis of species extinction owing to habitat loss and fragmentation (thus complementing Part I of the book). The process of diffusion as the basic way to describe population dispersal is initially illustrated. Demography is then added to diffusion to describe the invasion of a species, thus leading to the reaction-diffusion model. In this way, one can estimate not only the impact of the spread of unwanted and alien species but also the speed of recolonization of space by a threatened species that has been rescued via suitable conservation policies. Another product of the model is the estimation of the critical dimension of reserves set up for biodiversity protection. The process of habitat loss and fragmentation is then modeled via the metapopulation approach devised by Richard Levins. His simple model, which includes also the impact of environmental catastrophes, is complemented by a spatially explicit approach leading to the concept of metapopulation capacity.

Part III deals with the problem of sustainable harvesting of biological resources (such as forests and fish stocks). The basic problem of renewable resources management (the tragedy of the commons) is introduced by illustrating the open access nature of these resources and the consequences of not regulating their exploitation. Different management goals are taken into account: the maximization of harvested biomass, the minimization of extinction risk, and the maximization of sustainable monetary benefit including the cost of harvesting effort. These goals can be achieved by different policies, both exclusive and nonexclusive, which are here described. Production curves relating effort to biomass yield are obtained to introduce the concept of Maximum Sustainable Yield (MSY). This simple biological approach is then complemented by principles of bioeconomics according to H. Scott Gordon's analysis: bionomic equilibrium and the effect of opportunity cost. The socioeconomic impacts of various regulation policies are discussed. The problem of discounting economic benefits is then illustrated by the analysis of the optimal rotation period in a managed forest or aquaculture. Beverton and Holt's theory for the optimal harvesting of constant-recruitment fish populations concludes this part.

Part IV of the book illustrates the basics of the modern theory on the ecology of parasites and diseases. It clarifies the profound link between ecology and public health illustrating the problem of emerging and reemerging diseases, chiefly zoonoses related to the overexploitation of our environment. The traditional division between micro and macroparasitic diseases is used. Simple SI and SIR models are illustrated and the fundamental concept of basic reproduction number is then introduced. Specific models for vector-borne (Ross-Macdonald) and water-borne diseases are described, while macroparasitic pathologies are modeled by the classical approach of Anderson and May which accounts for the distribution of parasite burden inside a host. A brief hint to the dynamics of parasitoids used for biological control concludes the book.

A unique feature of this book is the large panoply of numerical problems proposed at the end of each part. The majority is based on real data derived from the ecological literature. The interested reader can thus practice the modeling tools previously illustrated in the various chapters.

Milan, Italy   Marino Gatto
February 2022   Renato Casagrandi

Acknowledgements

This book originates from the lectures of the graduate courses of Ecology we have been holding in the past twenty years at Politecnico di Milano, our home institution to which we are indebted for the help that has always been provided to us. In particular, our department (Dipartimento di Elettronica, Informazione e Bioingegneria) is a wonderful academic environment that systematically offers us the chance to work at our best and achieve the goals we have in mind. Gratitude must first be expressed to scientists who have been our masters in either direct (*first-hand*) or indirect ways (having been masters of either Marino or Renato): Colin Clark, Buzz Holling, Simon Levin, Carlo Matessi, Sergio Rinaldi, Larry Slobodkin, and Carl Walters. Their teaching has been fundamental for the development of many of the ideas present in the book. We acknowledge also many of our colleagues and friends who have been generous in help and advice throughout the years and whose work has been an inspiration for us. To just name a few: Enrico Bertuzzo, Isabella Cattadori, Giulio De Leo, Giorgio Guariso, Ran Nathan, Carlo Piccardi, Andrea Pugliese, Andrea Rinaldo, Ignacio Rodriguez-Iturbe, and Nicola Saino. A special mention and gratitude in this regard go to our colleagues of the Ecology Group at Politecnico di Milano, Lorenzo Mari, and Paco Melià with whom we share an untamed passion for research and teaching. Not to be forgotten is the enthusiasm of our undergraduate, graduate students, and postdocs in Ecology that has been a constant and great stimulus for us both. In particular, we would like to mention Daniele Bevacqua, Luca Bolzoni, Lorenzo Righetto, Manuela Ciddio, Andrea Mignatti, Mattia Pancerasa, Francesca Recanati, Francesca Dagostin, Chiara Vanalli, Alba Bernini, Giulia Fiorese, Marcello Schiavina, Amadou Lamine Toure, and Federica Guerrini. Finally, our deepest thanks go to Claudia and Elena for their continuous support throughout the past years, independent of all difficulties including those induced by the COVID-19 pandemic. Without their love (and patience!), we would have not been able to complete this work.

Milan, Italy
February 2022

Marino Gatto
Renato Casagrandi

Contents

Part I
Species and Populations Threatened by Extinction

Chapter 1
Threatened Biodiversity

1.1 What is Biodiversity?

The impressive diversity of life forms on the planet Earth and their distribution over the globe have always aroused the wonder and curiosity of many scientists and enthusiast naturalists. The term biodiversity, or biological diversity, indicates precisely all of these life forms. The history of biodiversity coincides with the history of life on Earth, life that originated in all probability about 3.7 billion years ago, certainly in the water, perhaps in deep hot springs (see Fig. 1.1). We must not forget that the earth's crust and atmosphere, as we know them now, have been moulded by the evolving organisms. The first living beings inhabited anoxic environments (originally there was virtually no oxygen in both water and the atmosphere). The evolution of organisms capable of performing photosynthesis allowed the accumulation of oxygen and the development of an atmosphere like the one we experience today.

From the very moment life appeared on our planet until now new organisms have been evolving, thus producing an incredible branching and diversification of life forms as demonstrated by systems of taxonomic classification (Fig. 1.2). In his book *"What is life?"*, the Nobel Prize for Physics Erwin Schrödinger (1944) had predicted that the key part of a living cell was to be an *aperiodic crystal*. His prophecy, that in some way was a source of inspiration for James Watson (at least according to what he states, Watson and Berry 2003), was confirmed by the discovery of the structure of DNA and other nucleic acids a few years later as a result of the work of Crick and the same Watson, who were however relying on the discoveries of the late Rosalind Franklin. Schrödinger pointed out that the genome must be on the one hand very stable (after all, one of the main characteristics of the organisms is that they can make copies of themselves), but on the other hand quite flexible, i.e., capable of development and evolution. In fact, the structure of the DNA possesses exactly these remarkable characteristics.

Without knowing anything about genetics and molecular biology, one of the giants of modern scientific thinking, Charles Darwin, had already understood the main mechanisms of this fantastic ramification (Darwin 1859). As we will see later, the

© The Author(s), under exclusive license to Springer Nature Switzerland AG 2022
M. Gatto and R. Casagrandi, *Ecosystem Conservation and Management*,
https://doi.org/10.1007/978-3-031-09480-4_1

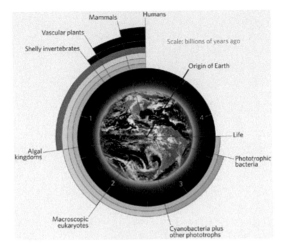

Fig. 1.1 A diagrammatic clock which summarizes the various eras of life on Earth (Des Marais 2005)

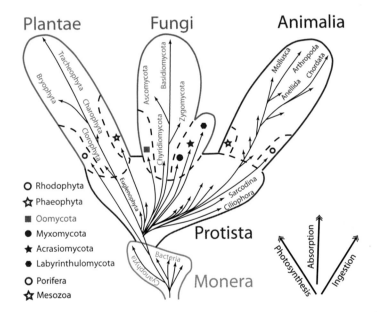

Fig. 1.2 The branching and diversity of life as described by the classification into five kingdoms of Whittaker (redrawn and simplified after Whittaker 1969). This system is based on three levels of organization: prokaryotes (kingdom of Monera), unicellular eukaryotic organisms (kingdom of protists) and multicellular eukaryotic organisms (kingdoms of plants, fungi and animals). The last three kingdoms are distinguished essentially by the differences in the acquisition of vital resources, as indicated symbolically in the lower right corner. Some *phyla* are difficult to classify: *Rhodophyta* (red algae) and *Phaeophyta* (brown algae) are traditionally classified as protists, *Oomycota* (water molds) as fungi, but *Myxomycota* and *Acrasiomycota* (mucilaginous fungi) and *Labyrinthulomycota* as protists, *Porifera* (sponges) and *Mesozoa* (worm-like parasites of marine invertebrates) as animals

phenomenon of genetic mutation is the main driver of the diversification of life, but if mutation, which is basically random, were the sole cause of diversification, we would observe no organized and coherent structures in the terrestrial biosphere. Instead, the organization of life is set up by the process of natural selection. In fact, in the first place, mutations are very rare, because of the genome stability, and many mutations are also deleterious. Of those not detrimental to the survival of organisms only a few manage to pass through the filter of natural selection, which favours those mutants that have a demographic advantage. Selection is very strict, but the filter operates continuously on a myriad of organisms; therefore it is very effective in the long run and produces organisms adapted to their environment. In small populations, neutral mutations may also establish due to a random process known as genetic drift (this fact was pointed out in more recenttime by Kimura and Crow 1964). Also, we must not forget that during the branching of life the evolution of the various organisms has been constrained by the presence of other organisms with which they have been interacting ecologically: therefore, evolution is in fact co-evolution and adaptation is indeed co-adaptation.

But genetics and taxonomy are not sufficient to fully define biodiversity. Actually, we can distinguish hierarchies of biological diversity. There exists diversity of genomes within a species, diversity of species or other taxonomic classes within an ecosystem and diversity of ecosystems within a landscape or biome. Traditionally, the diversity of species within an ecosystem (also called α-diversity) is the concept most commonly used, but there is a growing consensus among biologists that diversity of the functions performed by species within an ecosystem and diversity of habitats in a landscape play a very important role and deserve much more attention than the one which has been paid to them until now (to this purpose, it is possible to introduce other dimensions of ecological diversity, such as the β and γ-diversity (Whittaker 1960). If we consider a single ecosystem, we can, for example, pay attention to its trophic structure distinguishing primary producers (plants), herbivores, carnivores and decomposers. Very often, it is reasonable to study the diversity of species within each trophic level or of each functional group of organisms that play a similar role within the ecosystem. However, the overall design of the ecosystem diversity can be captured only by coupling these studies with the analysis of the food web structure and of the flows of energy and matter within that network.

1.2   Configuration and Distribution of Biodiversity

The simplest question we can ask is: how much do we know, in purely descriptive terms, about biodiversity on our planet? The most honest answer is "very little ". We do not even know how many species are hosted by our mother earth. The species described so far are almost 2 million (Purvis and Hector 2000), but it is estimated that the total number of species on the planet is much larger. Different methods, valid primarily for terrestrial ecosystems, lead to the estimation that the number might be between 3 and 30 million. The most trusted value is 14 million (Hawksworth

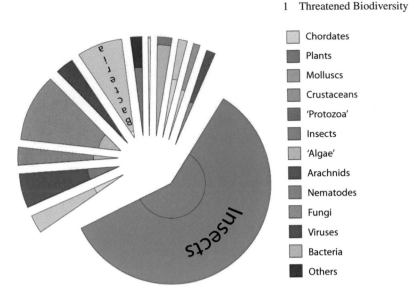

Fig. 1.3 The way biodiversity of the earth is probably divided among different taxonomic groups (after Purvis and Hector 2000). The area of each slice of the pie represents the number of species that is estimated to exist within each group; interior sectors represent the number of those species that have actually been catalogued

and Kalin-Arroyo 1995). In particular, marine ecosystems are very little known. The marine species described so far are about 250,000, but recent studies (Bouchet 2006) estimate that the oceans could accommodate up to 10 million species. In addition, the knowledge of the world biodiversity is not the same for the different taxonomic groups (see Fig. 1.3). It is very good as far as birds and mammals are concerned, but, for example, it is very bad with respect to bacteria. Insects are the true dominators of the earth because they have branched into an incredible number of species.

Biodiversity is not evenly distributed on Earth. There are some regions that contain a very large number of different species that are endemic, namely present only in that particular geographical location. These regions (see Fig. 1.4) were called *hotspots* of biodiversity by Norman Myers in an article which is fundamental for conservation ecology (Myers et al. 2000). Of course, hotspots are very important because they require special attention by the global policies that are proposed to safeguard our planet's biodiversity. It is important to remark that Italy, the country of this book's authors, with its great diversity of landscapes and ecosystems, is one of these hotspots, thus deserving appropriate conservation policies and management of its biological heritage.

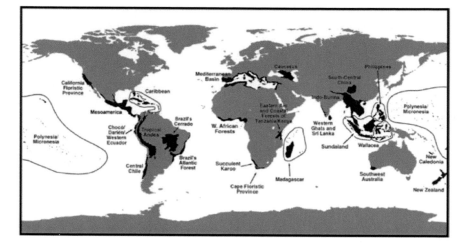

Fig. 1.4 The *hotspots* of biodiversity (Myers et al. 2000). They are defined as "areas that contain exceptional concentrations of endemic species and that are subject to an exceptional loss of habitat"

1.3 The Loss of Biodiversity

The problem of present and future biodiversity loss is often confused by some people (even by some scientists) with the "natural" decreases of biodiversity that occurred over evolutionary time scales. Our planet has always experienced losses of biodiversity from the very beginning of life on earth. To be precise, over a geological scale there have been five periods of mass extinction, termed according to the corresponding geological era, followed by radiation, namely the appearance of new species. Figure 1.5 shows the change of the number of taxonomic classes in the last 500 million years (revised after Primack 2000). It is apparent that the number of taxonomic classes has been increasing over time, even if the increase has not been regular. Currently, we are probably at the beginning of a sixth extinction, that of the Holocene. The causes of past extinctions are not very well-known. Instead, the cause of the forthcoming sixth extinction is quite well-known because it is the action of man. That's why Paul Crutzen (2002) has termed the present era the *Anthropocene*.

When skeptics argue that species have always become extinct and that we should not worry about this phenomenon, they make a very common mistake: they do not pay attention to the time and space scales. In fact, while the geological average lifetime of a species was of the order of a few million years, the current average residual time of many species is estimated to be of the order of a few tens of thousands of years. Figure 1.6 shows the changes over time of the extinction rates of mammals and birds, as estimated by Smith et al. (1993). Note that these two taxa are indeed at risk, because their the average lifetime is reduced to a few hundred years (Lawton and May 1995). The fact that the extinction rates have been very low in very recent years (see again Fig. 1.6) is not due, as one might erroneously think, to the positive impact

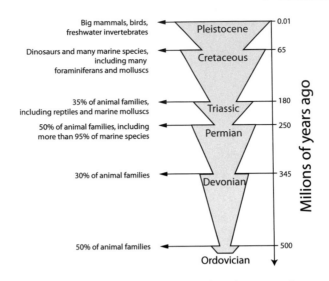

Fig. 1.5 A much simplified scheme of the great extinctions that occurred over geological time (revised after Primack 2000). The width of the gray area indicates the relative abundance of living taxa, the name of the period or geological era in which the mass extinction occurred is indicated by horizontal arrows, to the right of which are listed the major taxonomic groups that suffered mass extinctions. The numerical labels shown on the vertical axis on the right indicate the distance in time of the extinction occurrence from the present era (in millions of years)

Fig. 1.6 Rates of species extinctions that occurred since 1600, reworked after Smith et al. (1993)

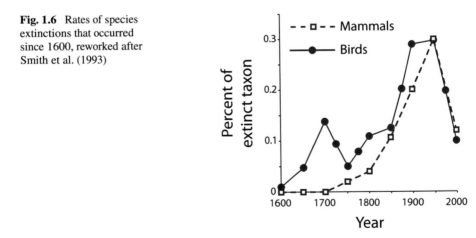

of conservation policies. Rather, it is the criteria that are now adopted to declare a species as extinct that have become stricter. Also, there are now technologies that allow the recovery and monitoring of even a very small number of individuals of a declining species. This explains a recent decrease of the number of species that are actually declared extinct (see below for the different definitions of extinction).

Table 1.1 Total number of extinctions recorded from 1600 to present time (fifth column) divided by taxon (first column) and habitat (from the second to the fourth column). The last column shows the percentage of the taxon that have become extinct with respect to the estimated total, while the sixth column shows the percentage of extinctions that have occurred on islands. Data from Primack (2000)

Taxa	Number of extinctions in				% Islands	% Total
	Islands	Land	Ocean	Total		
Mammals	51	30	4	85	60	2.1
Birds	92	21	0	113	81	1.3
Reptiles	20	1	0	21	95	0.3
Amphibians	0	2	0	0	2	0.05
Fish	1	22	0	21	4	0.1
Invertebrates	48	49	1	98	49	0.01
Flowering plants	139	245	0	384	36	0.2

As results from Table 1.1, islands are particularly vulnerable environments. Remember, in this regard, the relationship between number of species and area illustrated in the theory of island biogeography (MacArthur and Wilson 1967). For example, of the many species of birds currently hosted in Hawaii there are very few that are indigenous because most of them have become extinct or are on the brink of extinction. Of the existing 272 species found in Hawaii, 54 are alien, 50 species are indigenous residents, 155 are migratory species that do not reproduce in Hawaii and 13 are migratory species that breed in the archipelago. It is estimated that before the arrival of man (about 2,000 years ago) there were at least 128 species of native nesting birds (The American Ornithologists' Union 1983). Usually, when we talk about extinct or endangered species we think of charismatic exotic species of unquestioned charm such as the black rhino, the African elephant, the giant panda, the dodo. However, even in Italy there exist many species that are under threat or have been in the past. Among these it is worth remembering the best-known cases: the wolf, the lynx, the brown bear, the ibex, the Sardinian deer, the monk seal, the otter, the golden eagle, the bearded and the griffon vultures, the capercaillie, the grey partridge. Other examples of some lesser known species that are crtitically endangered or definitively extinct in Italy are shown in Fig. 1.7.

It is important to remark that there are different levels of endangerment or extinction threat, levels that are now internationally recognized by the IUCN (International Union for Conservation of Nature and Natural Resources, http://www.iucnredlist.org/). This institution regularly produces *red lists*, i.e. lists of endangered animal, fungi and plant species. A species is considered *extinct* when all over the world there no longer exists (as far as humans know) even a single living individual of that species. For example (see Fig. 1.8) Bachman's warbler *Vermivora bachmanii* is considered extinct. If a species is still represented by a few individuals living in a zoo or in a botanical garden, or more in general in captivity, then the species is termed *extinct in the wild*. If the extinction refers to the whole biosphere, species

Fig. 1.7 Some of the animal and plant species that are critically endangered (like *Anchusa littorea*), see (Fenu et al. (2013)) or extinct in Italy (like *Aldrovanda vesiculosa*, *Haliaeetus albicilla* and *Oxyura leucocephala*)

are thus said to be *globally extinct*. In some cases, however, the extinction did not occur throughout the world, but only in some regions. In this case the species are termed *locally extinct*. For example (see again Fig. 1.8) the American burying beetle (*Nicrophorus americanus*) was widespread in the eastern and western United States (inset map), while today only three populations of this species persist, two in central US and one in Long Island, New York. Even the white-headed duck (*Oxyura leucocephala*, see Fig. 1.7), a small diving duck, is in these same conditions: extinct in all of Eastern Europe, in Italy and Corsica, it still existed in the 1970s in Spain, where a small population was still thriving, was fortunately placed under protection and is now recovering. In some cases, unfortunately, these small local populations are almost certainly doomed to extinction. For example, in Somalia there exist only two trees of *Moringa pygmaea*, real "Living Dead" (see again Fig. 1.8): when they

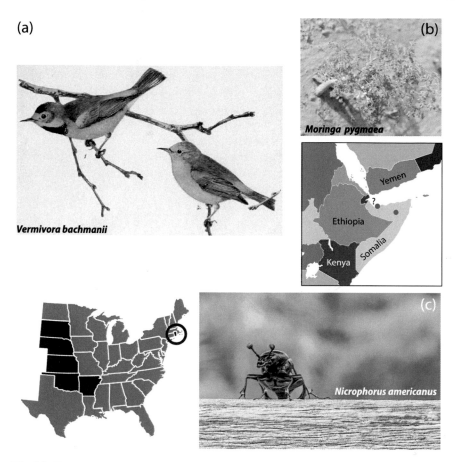

Fig. 1.8 Examples of different extinction levels. **a** A print after Chapman (1907) depicting an adult male (left) and an adult female (right) of Bachman's warbler (*Vermivora bachmanii*) a globally extinct species because of deforestation. **b** One of the two specimens found in Somalia (green circled locations) in 2001 of the species *Moringa pygmaea*, which is considered a "living dead" (photo courtesy of Prof. Mats Thuin). **c** Once widespread throughout the eastern and central United States, the American burying beetle (*Nicrophorus americanus*) is confined to a very small territory, being locally extinct in many areas it occupied

die the species will become extinct. In the international literature species of this sort are termed *dead-living species*. Finally, some ecologists call *ecologically extinct* the species that are able to persist even with so small populations that their ecological role is practically negligible. An example of ecologically extinct species is the Asian tiger, whose effect on prey populations is now basically immaterial.

1.4   Causes of Threat and Extinction

What are the major factors that lead the species to extinction? Several statistical analyses, though approximate, were made. The main cause of the extinctions occurred so far is the change of land use, which leads to the degradation and destruction of habitats. Additional causes are the introduction of alien species, the harvest of animals and plants and pollution. Figure 1.9 shows for example the main causes of extinction and threat to the bird populations at the global level.

There are many examples of habitat degradation, but nowadays the most impressive is the destruction of tropical forests. While Europe's forests have been destroyed and fragmented in the course of centuries and with the aim of building towns and cities that are part of our culture, tropical forests (see Fig. 1.10) are destroyed over much shorter time scales (tens of years) and basically in order to produce resources for what we call the developed countries. Often the fragmentation is linked to the construction of roads as is the case for the Trans-Amazonian Highway road network. Sometimes there is a conversion of land use in favour of agricultural crops such as soy-bean that are primarily intended for export. This conversion is financed by international loans. The destruction is unfortunately very fast and creates an unnatural mosaic of fragments at different territorial scales. Figure 1.11 shows an example of

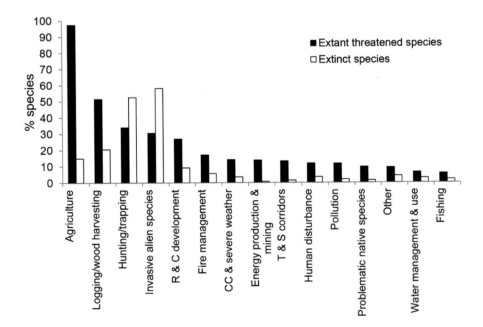

Fig. 1.9 Percentage importance of the causes of extinction of and threat to bird species worldwide (Szabo et al. 2012). Abbreviations: R & C stays for Residential and Commercial, CC for Climate Change and T & S for Transport and Services

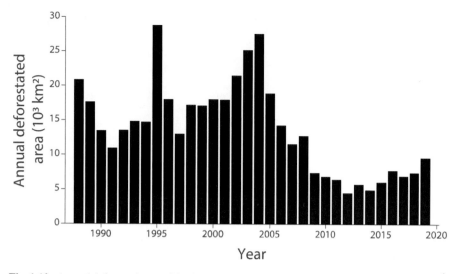

Fig. 1.10 Annual deforested area of the Amazon forest in the last three decades (thousands of km^2)

a Landsat7 image taken in August 2000 that frames the new agricultural settlements east of Santa Cruz de la Sierra, Bolivia, in an area of tropical dry forest.

Many models have been proposed to understand and predict the loss of biodiversity due to habitat fragmentation. To better define the problem, it is worthwhile to remark that even undisturbed natural populations are in many cases divided into subpopulations. The butterflies *Melitaea cinxia* of Ålan Islands in Finland (Hanski et al. 1995) and *Euphydryas editha bayensis* in California (Harrison et al. 1988) are two famous examples of this kind. The first species lives in a landscape made up of hundreds of pieces of vegetation located in many islands of the Ålan archipelago, while the second requires a particular habitat (grasslands with serpentine soil) which is patchily distributed in California. As we will better specify in Chap. 6, the famous ecologist (Levins 1970) was the first to coin the term *metapopulation*, or "population of populations", to indicate a set of local populations where local extinctions are balanced by immigration due to emigration from other subpopulations that are present in fragments of still occupied habitat. The metapopulation is the current paradigm not only for populations inhabiting naturally inhomogeneous landscapes but also for populations threatened by habitat fragmentation.

A major extinction driver, though little known to the general public, is the introduction of alien species. These have not evolved in the environment in which they are introduced, therefore they neither are adapted to it nor co-evolved together with the indigenous species. They often have a competitive advantage over native species, or carry parasites and diseases to which native species have not adapted, or act as new predators against which native species are defenceless. Examples abound. The grey squirrel (*Sciurus carolinensis*) was introduced from North America to Britain in late 1800. It has few natural predators and is bigger and more athletic than the

Fig. 1.11 a Landsat7 image, taken on August 1, 2000, showing new agricultural settlements east of Santa Cruz de la Sierra, Bolivia, in an area of tropical forest. Since the 1980s, this region has been rapidly deforested as a result of a re-settlement from the Andes highlands and an extensive agricultural development project called Tierras Baja. **b** The radial pattern of fields is part of the re-settlement of San Javier. At the center of the settlement there is a small community which includes a church, a coffee bar, a school and a soccer field—the essentials of rural life in Bolivia. **c** Straightlined, clear-colored areas are soybean fields cultivated for export and mainly financed by foreign capital. The dark stripes across the fields are windbreaks which are necessary because the soils of these areas are fine-textured and subject to erosion. Credit: https://earthobservatory.nasa. gov/images/1053/deforestation-in-bolivia

Mnemiopsis leidyi   **Lates niloticus**

Fig. 1.12 Examples of invasive alien species discussed in the text: the comb jelly *Mnemiopsis leidyi* and the Nile perch *Lates niloticus* of the Lake Victoria in comparison with a man (image credit: user smudger888 on Flickr)

native species, i.e. the red squirrel (*Sciurus vulgaris*). Additionally, it is the vector of *Parapoxvirus*, which is lethal to the red squirrel, and better resists the fragmentation of forest habitat. The American species has replaced the European one almost anywhere in Britain. The distribution area of the red squirrel is now reduced to the island of Anglesey and the North of Scotland. The comb jelly *Mnemiopsis leidyi*, a marine ctenophore (see Fig. 1.12), native to the western Atlantic coast, was accidentally introduced into the Black Sea in 1980 probably through the ballast waters of merchant ships. Here it greatly multiplied; as *M. leidyi* is a carnivore that feeds on eggs and larvae of pelagic fish, it has caused a dramatic decrease of the harvest of Black Sea fisheries, especially of the anchovy (*Engraulis encrasicholus*). The Nile perch (*Lates niloticus*, see Fig. 1.12) was introduced into Lake Victoria in East-Central Africa to give a boost to the local fishing industry. Being a large predator it has caused the extinction or near extinction of more than 200 species of cichlid fish that are found only in the largest African lake.

A further cause of threat to many species lies in overexploitation due to fishing and hunting. The harvest can be the primary cause or aggravate situations already at risk because of habitat degradation. The species most threatened by hunting and fishing are not only those whose flesh is edible (typically game and fish stocks), but also those whose skin and whose horns, tissues and organs have a high commercial value (like the elephant tusks, from which humans obtain ivory, or the rhino horn to which aphrodisiac properties are falsely ascribed to). Hunting and fishing do not always affect the ecosystem diversity, but become a serious extinction threat for species that are exploited excessively, i.e. whenever the catch rate is larger than the renewal rate of the species biomass or numbers.

Great importance as an extinction or threat cause can be attributed to pollution. Human activities, in fact, have profoundly altered the biogeochemical cycles which are critical to the overall functioning of ecosystems. Sources of pollution are, in addition to industries and civil waste, also the agricultural activities that utilize insecticides, pesticides and herbicides, thus profoundly altering the soil and the food chains. It is worthwhile to remind the reader about the phenomenon of biomagnification, which consists in the amplification of the concentration of toxic substances in the food webs from the lower levels (primary producers) to the highest levels (top predators). The consequence of this process is the accumulation of significant quantities of harmful chemicals (such as heavy metals) in organisms that are at the top of the food chain (raptors, large carnivores).

In addition to analysing the reasons for the loss of biodiversity in the past we can also wonder what causes will mainly drive the loss of biodiversity in the future. According to Sala et al. (2000) the major impacts on biodiversity between now and 2100 will be those listed and compared in Fig. 1.13. The main cause will still be land-use change, leading to the degradation and destruction of habitats, followed by global climate change (GCC), the alteration of the nitrogen cycle, the introduction of alien species, the increase of carbon dioxide (CO_2). If we consider that global climate change is mainly due to rising concentrations of carbon dioxide and methane, we can state that the increase in greenhouse gases will therefore be in the future the most important cause of biodiversity loss, followed by land-use change on land and by human pressure (fishing and coastal urbanization) in the marine environment. The effects of GCC are now particularly severe in the arctic, boreal and alpine biomes, where climate is the main factor regulating the ecosystems, but will be increasingly important in changing ecosystems functioning almost everywhere in the next 50 years and beyond (IPCC-WGII 2014).

The golden toad (*Bufo periglenes*) in the mountains of Costa Rica is probably the first species that was officially declared extinct because of GCC. It is estimated that 67% of the 110 species of harlequin frogs (*Atelopus* sp.), which are endemic to the American tropics, are going to meet the same fate (Pounds et al. 2006). Other species have escaped GCC moving towards the poles or to higher altitudes. Still others, with possible effects on human health, such as *Anopheles* mosquitoes that carry malaria, will find more favourable conditions in warmer environments. In oceans the coral reefs are the ecosystems that harbour the majority of biodiversity. They are very sensitive to temperature because corals can survive only within a narrow temperature range. Several episodes of coral bleaching have already been occurring, and they are certainly linked to the increasing temperature of ocean waters as documented by NOAA-NESDIS (*National Oceanic and Atmospheric Administration, National Environmental Satellite Data and Information Service*) and UNEP-WCMC (*United Nations Environment Programme—World Conservation Monitoring Centre*).

It should be noted that organisms are among the best indicators of global warming. The "ecological footprints" of GCC, as they are called, are very clear. Very apparent is the phenological shift of plants and animals (typically, earlier breeding and flowering

Fig. 1.13 Estimate for the year 2100 of the relative effect on biodiversity of the main drivers of change; the thin bars represent the variability of the impact in the different terrestrial biomes (after Sala et al. 2000)

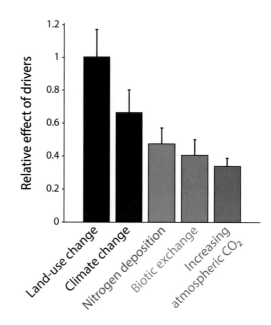

seasons). In recent decades, the average shift has been five days per decade and the birds were the most sensitive to the climatic signal (Root et al. 2003). Researchers have also found a shift towards the poles of indicator species such as butterflies (Parmesan et al. 1999). Very clear is also the shift towards higher altitudes. For example, the response to global warming of alpine vegetation in Europe has been a shift upwards of several metres per decade (Grabherr et al. 1994).

1.5 The Value of Biodiversity

A question that is often asked concerns the reasons why we should preserve biodiversity: Is it really so valuable? Isn't it perhaps a waste of money, a luxury that only the affluent countries can afford? Of course, answering this kind of questions requires not only scientific expertise, but also socio-economic and even ethical considerations. The social sciences distinguish between *direct value* (what we can get directly from organisms, that is food, fiber, timber, etc.), *indirect value* (related to the services that ecosystems render to humanity for free) and *intrinsic value* (aesthetic, spiritual, etc.).

While the direct value of biodiversity needs no special comments, it is good to consider indirect value in more depth. We begin by giving a very clear example of the value of ecosystem services. During the tsunami that struck East Asia in December 2004, the areas of Sri Lanka where the coral reef was still intact were invaded by the huge wave only 50 m from the shoreline and no death was recorded. Unfortunately, this did not happen a few kilometres away where the reefs had been illegally destroyed (Fernando et al. 2005). Here the dead were counted in hundreds.

The services provided by ecosystems are innumerable. The following list of benefits is certainly incomplete, but can serve as a guide for further consideration:

- ecosystems moderate extreme weather events, thus contributing to climate stability;
- mitigate the impacts of floods and droughts;
- protect waterways and coastlines from erosion;
- protect people from harmful ultraviolet rays;
- recycle nutrient salts;
- purify the air and water;
- decompose and purify wastes;
- control agricultural pests;
- regulate organisms that are vectors of infectious diseases;
- disperse seeds and pollinate agricultural and natural vegetation;
- generate and preserve soils and regenerate their fertility.

While in the early 1970s there was a growing awareness that environmental assessment was as necessary as economic evaluation (just think of the legislation on *EIA*, the Environmental Impact Assessment, which was introduced in North America and Europe), in more recent times the old logic that economic evaluation should prevail over all other considerations gained new force. This new context led Costanza et al. (1997) to conduct a purely monetary, comprehensive evaluation study on the resources and services that biodiversity provides us with. The authors estimated that these services were worth 33 trillion dollars (to provide the reader with an idea of this enormous sum of money, consider that the whole earth's GDP, gross domestic product, was then worth 18 trillion dollars). Although this approach was criticized (see for example Gatto and De Leo 2000), one must admit that these results should convince even the most sceptical economist on the importance of preserving biodiversity. In other words, Costanza et al. (1997) showed that conserving biodiversity is paid back, because ecosystem services are worth at least twice the global GDP. It should be emphasized that any time we lose a species we might have lost forever, for example, the possibility of treating one of the diseases that ravage humanity. In fact 80% of the human population relies for its treatment on natural medicinal products. Of the 150 drugs most prescribed in the United States, 118 are derived from natural organisms: 74% from plants, 18% from mushrooms, 5% from bacteria and 3% from a vertebrate (snake). Nine of the 10 most used drugs are derived from natural plant products. More than 100,000 different species of animals, including bats, bees, flies, butterflies and birds, freely provide their services as pollinators. A third of the food consumed by humans comes from naturally pollinated plants. The value of natural pollination services in the US alone was estimated to be 4 to 6 billion US dollars a year. Several cost-benefit analyses were conducted regarding the services of ecosystems and all of them showed that it is not true that biodiversity conservation is too expensive for the developing world countries. Actually, preservation costs are largely offset by benefits (Balmford et al. 2002, 2003).

1.6 Preserving Biodiversity

An extremely difficult, but also very urgent, question we can ask at this point is what humanity should do to preserve the diversity of our biosphere. Obviously, we cannot expect to fully answer the question in a few lines. However, we try to introduce the reader to the problem by summarizing the terms of the most urgent question that arises nowadays: in what direction should we direct our efforts at the global level? As with all the other problems faced by humanity, the financial or other resources that can be devoted to solving the issue are necessarily finite. Therefore, there exists a problem of optimal allocation of money, time, labour and so on. "Where should our money go? Into preserving the Mediterranean flora or the Australian fauna?" As already stated at the beginning of this chapter, Myers et al. (2000) have identified 25 *hotspots* of biodiversity all over the world using as indicators the species endemism and the level of threat that species are subjected to. The analysis of Myers sparked a useful debate. Obviously, the optimal allocation of resources to conserve biodiversity is a multi-criterial problem: the answer depends on the criterion that is used to define optimality. Recently, Orme et al. (2005) conducted a very interesting analysis on the biodiversity of birds (avian fauna is very well-known) considering three different criteria for each site: species richness, the threat to the species, and the wealth of endemic species. The resulting hotspots are not the same, as the reader can see by examining the three maps in three different colours reported in Fig. 1.14. However, as noted by Gatto and Casagrandi (2007), there is a certain overlap between the areas identified by using the three different criteria. The wealth of endemic species seems to be the most representative criterion, because it includes areas and species that are also included by the other two criteria. It is remarkable that hotspots of avian fauna are mainly located in highlands and islands.

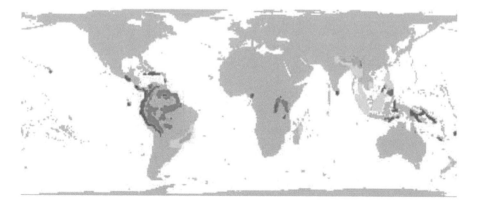

Fig. 1.14 Hotspots of avian species richness (in green), hotspots of endangered species (in yellow) and hotspots of endemic species (in red). Modified by Gatto and Casagrandi (2007) after data in Orme et al. (2005)

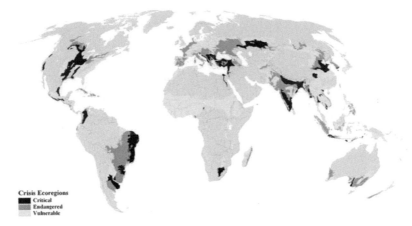

Fig. 1.15 Map of critically threatened ecoregions worldwide. Reproduced from Hoekstra et al. (2005)

The viewpoint of Myers et al. (2000) and Orme et al. (2005) has been partially contradicted by another analysis that considers a different indicator of ecological diversity. It is true that the species is the key concept of biodiversity and the basic unit of life evolution. However, as we already stated earlier, the functioning of ecosystems in their entirety is another extremely important issue. Even the aim of preserving individual species could be better achieved by aiming at the conservation not of the single species, rather of the overall habitat in which species live. Figure 1.15 (which is taken from Hoekstra et al. 2005) reports the map of ecoregions (defined as sets of ecological communities typically associated with specific geographic features) that are subject to global crisis. The map outlines another possible goal of ecology preservation, which is that of conserving habitats and biomes that are rapidly disappearing, such as the Mediterranean vegetation. Interestingly, the Tropical Andes (one of the *hotspots* of Myers) are no longer present in the map of Hoekstra et al. (2005). In fact, they have a very high richness of endemic species, but their habitat is not critically threatened as of now. On the other hand, we must consider that ecosystems that are home to low biodiversity such as, for example, vast tracts of boreal forest that stretch from Russia to Canada, provide important ecological services, for example the absorption of carbon from the atmosphere, which is crucial for mitigating global warming. These ecosystems were not classified by Myers and colleagues as hotspots of biodiversity, yet they are definitely worth being preserved. We can therefore say that the debate is still very open.

A very important issue for the conservation of global biodiversity is also that of gap species. During the *5th World Parks Congress, Durban, South Africa*, in September 2003, it was officially announced that the global network of protected areas covered at least formally, 11.5 % of the land of our planet. This seemed to be a good result, even beyond the objectives set ten years earlier. However, protected areas are not designed in a way to really preserve a considerable amount of biodiversity.

In particular, there are many species that have been officially declared endangered by IUCN and whose current range is not within any protected area (Rodrigues et al. 2004). These species have been named *gap* species. Overall, 20% of the endangered species that were analyzed were identified as gap species. In particular, among them there are many amphibians. It is therefore urgent to extend the network of protected areas in such a way that they host the greatest number of endangered or critically endangered species.

For more information on the currently most debated topics please refer to the 5th edition of *Global Biodiversity Outlook* (that can be freely downloaded at https://www.cbd.int/gbo5).

References

Balmford A, Bruner A, Cooper P, Costanza R, Farber S, Green R, Jenkins M, Jefferiss P, Jessamy V, Madden J, Munro K, Myers N, Naeem S, Paavola J, Rayment M, Rosendo S, Roughgarden J, Trumper K, Turner R (2002) Economic reasons for conserving wild nature. *Science* 297:950–953

Balmford A, Gaston KJ, Blyth S, James A, Kapos V (2003) Global variation in terrestrial conservation costs, conservation benefits, and unmet conservation needs. *PNAS* 100:1046–1050

Bouchet P (2006) The magnitude of marine biodiversity. In: *The Exploration of Marine Biodiversity. Scientific and Technical Challenges*, Fundación BBVA, pp 31–62

Chapman FM (1907) *Warblers of North America*. Appleton, New York, New York

Costanza R, d'Arge R, de Groot R, Farber S, Grasso M, Hannon B, Limburg K, Naeem S, O'Neill R, Paruelo J, Raskin R, Sutton P, Van den Belt M (1997) The value of the world's ecosystem services and natural capital. *Nature* 387:253–260

Crutzen PJ (2002) Geology of mankind. *Nature* 415:23

Darwin C (1859) *On the origin of species*. John Murray, London, U.K

Des Marais DJ (2005) Palaeobiology: sea change in sediments. *Nature* 437:826–827

Fenu G, Cogoni D, Ulian T, Bacchetta G (2013) The impact of human trampling on a threatened coastal mediterranean plant: the case of *Anchusa littorea* Moris (Boraginaceae). *Flora—Morphology, Distribution, Functional Ecology of Plants* 208(2):104–110

Fernando HJS, Mendis SG, McCulley JL, Perera K (2005) Coral poaching worsens tsunami destruction in Sri Lanka. *EOS, Transactions American Geophysical Union* 86:301–304

Gatto M, Casagrandi R (2007) Threatened biodiversity: understanding, predicting, taking action. In: Rodriguez-Iturbe I, Sanchez Sorondo M (eds) *Water and the Environment*. Pontificia Academia Scientiarum, Vatican City, pp 17–37

Gatto M, De Leo GA (2000) Pricing biodiversity and ecosystem services: the never ending story. *BioScience* 50:347–355

Grabherr G, Gottfried M, Pauli H (1994) Climate effects on mountain plants. *Nature* 369:448

Hanski I, Pakkala T, Kuussaari M, Lei G (1995) Metapopulation persistence of an endangered butterfly in a fragmented landscape. *Oikos* 72:21–28

Harrison S, Murphy DD, Ehrlich PR (1988) Distribution of the bay checkerspot butterfly, *Euphydryas editha bayensis*: evidence for a metapopulation model. *American Naturalist* 132:360–382

Hawksworth DL, Kalin-Arroyo MT (1995) Magnitude and distribution of biodiversity. In: Heywood VH (ed) *Global Biodiversity Assessment*. Cambridge University Press, Cambridge, U.K., pp 107–191

Hoekstra JM, Boucher TM, Ricketts TH, Roberts C (2005) Confronting a biome crisis: global disparities of habitat loss and protection. *Ecology Letters* 8:23–29

IPCC-WGII (2014) Ipcc, 2014: Summary for policymakers. In: Field C, Barros V, Dokken D, Mach
    K, Mastrandrea M, Bilir T, Chatterjee M, Ebi K, Estrada Y, Genova R, Girma B, Kissel E, Levy
    A, MacCracken S, Mastrandrea P, White L (eds) *Climate Change 2014: Impacts, Adaptation,
    and Vulnerability*. Cambridge University Press, Cambridge, United Kingdom and NewYork,
    NY, USA, pp 1–32, Contribution of Working Group II to the Fifth Assessment Report of the
    Intergovernmental Panel on Climate Change
Kimura M, Crow IF (1964) The number of alleles that can be maintained in a finite population.
    Genetics 49:725–738
Lawton JH, May RM (1995) *Extinction Rates*. Oxford University Press, Oxford, U.K
Levins R (1970) Extinction in some mathematical questions in biology. *Lecture Notes on Mathe-
    matics in the Life Sciences*. The American Mathematical Society, Providence, RI, pp 75–107
MacArthur RH, Wilson EO (1967) *The Theory of Island Biogeography*. Princeton University Press,
    Princeton, NJ, USA
Myers N, Mittermeier RA, Mittermeier CG, da Fonseca GAB, Kent J (2000) Biodiversity hotspots
    for conservation priorities. *Nature* 403:853–858
Orme CDL, Davies RG, Burgess M, Eigenbrod F, Pickup N, Olson VA, Webster AJ, Ding TS,
    Rasmussen PC, Ridgely RS, Stattersfield AJ, Bennett PM, Blackburn TM, Gaston KJ, Owens
    IPF, (2005) Global hotspots of species richness are not congruent with endemism or threat. *Nature*
    43:1016–1019
Parmesan C, Ryrholm N, Stefanescu C, Hill JK, Thomas CD, Descimon H, Huntley B, Kaila L, Kull-
    berg J, Tammaru T, Tennent WJ, Thomas JA, Warren M (1999) Poleward shifts in geographical
    ranges of butterfly species associated with regional warming. *Nature* 399:579–583
Pounds JA, Bustamante MR, Coloma LA, Consuegra JA, Fogden MPL, Foster PN, La Marca E,
    Masters KL, Merino-Viteri A, Puschendorf R, Ron SR, Sánchez-Azofeifa GA, Still CJ, Young
    BE (2006) Widespread amphibian extinctions from epidemic disease driven by global warming.
    Nature 439:161–167
Primack R (2000) *A Primer of Conservation Biology*. Sinauer Associates, Sunderland, MA, USA
Purvis A, Hector A (2000) Getting the measure of biodiversity. *Nature* 405:212–219
Rodrigues ASL, Andelman SJ, Bakarr MI, Boitani L, Brooks TM, Cowling RM, Fishpool LDC, da
    Fonseca GAB, Gaston KJ, Hoffmann M, Long JS, Marquet PA, Pilgrim JD, Pressey RL, Schipper
    J, Sechrest W, Stuart SN, Underhill LG, Waller RW, Watts MEJ, Yan X (2004) Effectiveness of
    the global protected area network in representing species diversity. *Nature* 428:640–643
Root T, Price JT, Hall KR, Schneider SH, Rosenzweig C, Pounds JA (2003) Fingerprints of global
    warming on wild animals and plants. *Nature* 421:57–60
Sala OE, Chapin FS III, Armesto J, Berlow E, Bloomfield J, Dirzo R, Huber-Sannwald E, Hueneke
    LF, Jackson RB, Kinzig A et al (2000) Global biodiversity scenarios for the year 2100. *Science*
    287:1770–1774
Schrödinger E (1944) *What is life?* Cambridge University Press, Cambridge, UK
Smith FDM, May RM, Pellew R, Johnson TH, Walter KR (1993) How much do we know about
    current extinction rate? *Trends in Ecology and Evolution* 8:375–378
Szabo JK, Khwaja N, Garnett ST, Butchart SHM (2012) Global patterns and drivers of avian
    extinctions at the species and subspecies level. *PLoS ONE* 7(10):e47080
Union The American Ornithologists' (ed) (1983) *Check-list of North American birds*, 6th edn. AOU,
    Washington, DC
Watson JD, Berry A (2003) *DNA: The Secret of Life*. Knopf
Whittaker R (1960) Vegetation of the Siskiyou mountains, Oregon and California. *Ecological Mono-
    graphs* 10:1–67
Whittaker RH (1969) Evolution of diversity in plant communities. *Brookhaven Symp. Biol.* 22:178–
    96

Chapter 2
The Risk of Extinction: Allee Effect and Genetic Deterioration

2.1 Demographic Phenomena in Populations Threatened by Extinction

Sections 1.3 and 1.4 of Chap. 1 have evidenced the biodiversity loss that affects our planet. It is therefore important to study in greater detail the phenomena that increase the vulnerability and extinction risk of animal and plant populations subject to increasing human pressure. To this purpose, it is necessary to classify the different mechanisms of impact. In this and the next chapter we will study only the simplest phenomena, i.e. those that operate at the level of individual populations (sets of organisms of the same species belonging to a given ecosystem). These factors can be roughly classified in the following way:

- inverse density dependence (the so-called *Allee effect* or *depensation*) caused by intraspecific interactions other than competition, such as for example sociality;
- genetic deterioration due to e.g. *genetic drift* or *inbreeding* in small populations;
- stochasticity due to either the important action of random events in small populations (*demographic stochasticity*) or the variability of external conditions that affect demographic parameters (*environmental stochasticity*).

The first two factors are studied in this chapter, the third in the next chapter.

All of these phenomena do not have in general a large impact on very abundant populations, but they become fundamental in all those species that are represented by small numbers either for natural reasons (for example because they are high in the food chain, e.g. top predators) or because they have been severely depleted due to human pressure. The simplest models of population dynamics (which we will briefly recapitulate in the next section) are sometimes inadequate to properly describe these phenomena. It is therefore necessary to introduce new modelling tools with different functional forms. In particular, compared to the traditional viewpoint, we will be obliged to introduce a probabilistic description of some phenomena and consider organisms as discrete entities, thus giving up the traditional approximation that utilizes continuous variables to describe animal and plant abundance. The important

M. Gatto and R. Casagrandi, *Ecosystem Conservation and Management*,
https://doi.org/10.1007/978-3-031-09480-4_2

issues of species invasion and habitat destruction and fragmentation (which we have described in the introductory chapter) will instead be the topics dealt with in the two chapters on spatial ecology and metapopulations.

2.2 A Brief Recapitulation of Basic Population Dynamics

Here, we briefly summarize some basic notions of population ecology (for more details see Iannelli and Pugliese 2014) that allow us to build the quantitative tools for analysing the consequences of the impacts enumerated above. In particular, the basics that will be used in the sequel are the concepts of Malthusian population and of density dependent demography, as well as the distinction between populations with reproduction concentrated in time and populations that distribute progeny production over time without a specific reproductive season. To this regard remember that if we let t be the time and N the population size (i.e. the number of conspecific organisms in a given ecosystem) or its density (the number of organisms per unit area or volume), the main demographic models (without considering populations structured according to age, size or stage) are as follows:

Malthusian demography   In this case, we assume that the resources per capita are constant (though not necessarily abundant) independently of the population size N and that there is no direct intraspecific interaction (such as defense of a territory or competition for space). Therefore, mortality and natality can be considered constant.

As for populations with concentrated reproduction (t = discrete time) the simple Malthusian model is

$$N_{t+1} = \lambda N_t$$

where the parameter λ is constant and is called the finite rate of demographic increase. The solution to this equation, given the initial population size $N(0) = N_0$, is a simple geometric growth, namely $N_t = N_0 \lambda^t$. It is often convenient to use a logarithmic scale for the population size N, thus obtaining

$$\log N_t = t \log \lambda + \log N_0.$$

This stipulates a linear relationship between $\log N_t$ and the discrete time t. If data of population size are available for different times t (e.g., yearly counts of animals) it is thus possible to use standard linear regression to estimate $\log \lambda$, hence λ.

As for populations with continuous reproduction (t = continuous time), instead, we can state (from now on we will use a "dot" over a variable to indicate derivation with respect to time)

$$\frac{dN}{dt} = \dot{N} = rN$$

where r is also constant and is called the *per capita* instantaneous rate of increase. It is the difference between the instantaneous birth rate ν and the instantaneous death rate μ, which are also supposed to be constant. The solution to this equation, given the initial population size $N(0) = N_0$, is a simple exponential growth, namely $N(t) = N_0 \exp(rt)$. Logarithmic transformation yields

$$\log N_t = rt + \log N_0.$$

Again, standard linear regression can be used to estimate the rate of increase r. Note the analogy between $\log \lambda$ and r, but also remark that λ is dimensionless (a pure number) while the dimension of r (as well as ν and μ) is time^{-1}.

Density-dependent demography   Obviously, if the rate of increase is larger than 1 (discrete time) or positive (continuous time), a Malthusian population would grow exponentially without bounds. This implies that sooner or later the assumptions that subtend the Malthusian growth are violated because competition for resources or for space will take place whenever density increases. We briefly describe the most frequently models employed in this case.

As for populations with concentrated reproduction

$$N_{t+1} = \Lambda(N_t)N_t.$$

The finite rate of increase Λ is not constant, but depends on the size or density of the population (which is equal to the size of the population divided by the area or volume occupied by the same population). The best-known models are those of Beverton and Holt (1957), summarized by the equation

$$N_{t+1} = \frac{\lambda N_t}{1 + \alpha N_t},$$

and that of Ricker (1954) given by the equation

$$N_{t+1} = \lambda N_t \exp(-\beta N_t),$$

where λ, α and β are positive constants.

The Beverton-Holt model was derived by assuming that there exists strong competition for resources among youngsters. The survival to adulthood is decreasing in a hyperbolic way with N and the parameter α describes the strength of competition. Therefore, N_{t+1} is an increasing function of N_t that saturates to λ/α and whose derivative at $N_t = 0$ is λ. Both α and the intrinsic finite rate of increase λ are proportional to the fertility of the organisms considered by Beverton and Holt. In many fish species the fertility can be very high (of the order of hundred thousands of eggs), so that the asymptotic value λ/α is practically reached for relatively low values of N_t. This implies that in very fertile species the density of the next year $t + 1$ is practically independent of the density of year t, even if the latter is quite large.

The mechanism implied by the Ricker model is different. Here, there exists strong competition among parents for reproduction, implying an exponential decrease of fertility with parental density N_t (subsumed by the parameter β). Therefore, N_{t+1} is a unimodal function of N_t that tends to 0 for $N_t \to \infty$ and whose derivative at $N_t = 0$ is λ. In this case, competition is so strong that, for large densities, the population density of the next year is a decreasing function of the parental population. This phenomenon is termed *overcompensation*.

The analysis of the dynamical behaviour of discrete-time models can be conducted in a simple way by searching for possible equilibria, that is solutions to the equation $N_{t+1} = \Lambda(N_t)N_t$ that are non-negative and constant in time, that is $N_t = \bar{N} = $ constant for any t. These steady states can thus be found as solutions to the equation $N = \Lambda(N)N$. $N = 0$ (that is extinction) is always a steady state, the other equilibria are found by imposing $1 = \Lambda(N)$, namely the finite rate of increase is equal to unity. In the case of the Beverton-Holt model we obtain $\bar{N} = (\lambda - 1)/\alpha$, while in the case of the Ricker model we get $\bar{N} = \log \lambda / \beta$. It is however basic to establish whether the various equilibria are stable or unstable, that is if a small initial perturbation of the equilibrium will decrease or increase in time, respectively. A simple graphical way to study the stability, which is called cobweb plot (sometimes termed Moran diagram), consists of drawing the function curve $\Lambda(N)N$ in the plane $N_t - N_{t+1}$ together with the 45° line (which by definition intersects $\Lambda(N)N$ at $N = 0$ and where $\Lambda(N) = 1$, that is at the equilibria). Then one proceeds in the following way

1.   set the initial population $N_0 > 0$;
2.   calculate $N_1 = \Lambda(N_0)N_0$ on the function curve;
3.   plot horizontally across from this point to the diagonal line; in this way N_1 is referred back to the axis N_t;
4.   plot vertically from the point on the diagonal to the function curve;
5.   repeat as many times as needed.

The procedure is illustrated with reference to the Ricker model in Fig. 2.1 Instead, for populations in which reproduction is continuous in time the general model is

$$\dot{N} = R(N)N$$

in which the per capita instantaneous rate of increase R is not constant, but depends on the population density N. The most used model is the logistic one given by

$$\dot{N} = rN \left(1 - \frac{N}{K} \right)$$

where K is the carrying capacity. It can be derived by assuming that the death rate μ linearly increases with density N and/or the birth rate ν linearly decreases with density N. The solution of the logistic equation is

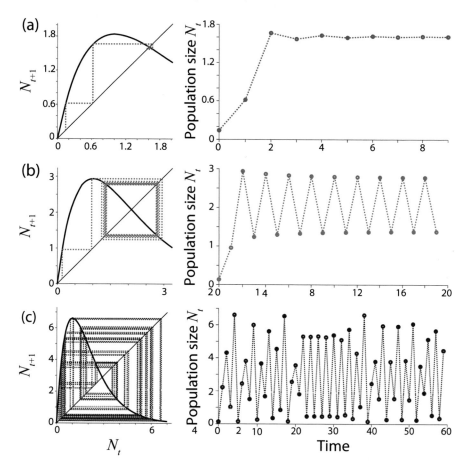

Fig. 2.1 Graphical analysis of the dynamical behavior of the Ricker model $N_{t+1} = \lambda N_t \exp(-\beta N_t)$ with $\beta = 1$ and λ equal respectively to **a** 5, **b** 8, and **c** 18; cobweb plots on the left, time dynamics on the right

$$N(t) = \frac{N(0)\exp(rt)}{1 + \frac{N(0)(\exp(rt)-1)}{K}}.$$

Therefore, given any initial condition $N(0)$ that is positive, the population density will tend towards the carrying capacity K.

An analytical way to ascertain whether equilibria (for both discrete and continuous-time models) are stable is via the linearization method, which consists in approximating the dynamics of a nonlinear system near each equilibrium with a linear system. This latter describes the dynamics of the small perturbations $\Delta N(t) = N(t) - \bar{N}$ around the equilibrium $\bar{N}$. In practice, as for discrete-time systems, one computes the derivative of $\Lambda(N)N$ at $\bar{N}$, namely $\Lambda'\left(\bar{N}\right)\bar{N} + \Lambda\left(\bar{N}\right)$, where we indicate the derivative of a function with respect to its generic variable (in this case N) by an apostrophe. If the modulus of the derivative $\left|\Lambda'\left(\bar{N}\right)\bar{N} + \Lambda\left(\bar{N}\right)\right| < 1$, then the equilibrium is stable. For continuous-time models, instead one computes the derivative of $R(N)N$ at $\bar{N}$, namely $R'\left(\bar{N}\right)\bar{N} + R\left(\bar{N}\right)$. If $R'\left(\bar{N}\right)\bar{N} + R\left(\bar{N}\right) < 0$, then the equilibrium is stable.

2.3 Depensation Phenomena and the Allee effect

One of the fundamental assumptions of the Malthusian population theory is that the ability of each individual to survive and/or reproduce is not affected by other individuals belonging to the same population. There is abundant experimental evidence to support the conclusion that in almost all populations this is not true. In most cases this mutual influence is negative, in the sense that it leads to a decrease of the ability of each organism to survive and reproduce (phenomenon of intraspecific competition). The parameter that plays a fundamental role is the density, i.e. the number of individuals in a certain area or volume: the bigger the density, the greater the mutual negative influence.

However, the presence of conspecific organisms is not always negative. In populations with a social structure, the inclusion of a further individual in a herd or flock can have a positive effect, because it can, for example, allow a better defence against predators or a more effective food research or a better offspring care. In this case there still exist density dependence effects, but they operate in the opposite direction, i.e. increasing density leads to higher birth rate and/or survival. This phenomenon is called *depensation* or *Allee effect*. Indeed, it was the American biologist Warner Allee (1931) the first to document an inverse dependence on density in the flour beetle *Tribolium confusum* (see Fig. 2.2). Subsequently, numerous examples have been well documented in natural populations outside the laboratory.

Fig. 2.2 Per capita reproduction rate evaluated over a period of 11 days (filled circles) or 25 days (open circles) in a population of the flour beetle *Tribolium confusum* (inset) as a function of the initial density of individuals. Redrawn after the original figure reported in Allee (1931)

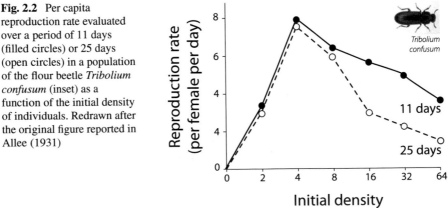

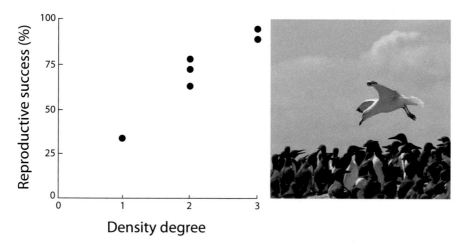

Fig. 2.3 The left panel (modified after Birkhead 1977) reports the positive influence of population density (assessed by a synthetic indicator) in the common guillemot *Uria aalge* on the percentage of individuals that manage to reproduce. The right panel shows a see-gull flying over guillemots

A first case of depensation in natural populations is that reported in the study of Birkhead (1977) on the common guillemot (*Uria aalge*) in Skomer island, South Wales. The left panel of Fig. 2.3 shows how the percentage of individuals that are able to reproduce, therefore the fertility, increases with density. This phenomenon is likely caused by the cooperation mechanisms against the attack of sea-gulls that prey on the eggs and juveniles of the guillemot (right panel of the same figure).

A second, very well documented example of Allee effect (Courchamp et al. 1999) is that pertaining to populations of wild dogs (*Lycaon pictus*, see Fig. 2.4) that have an important social structure. The wild dogs were once numerous and widespread in Africa, but now the wild dog is the most threatened large carnivore in this continent. Even in protected areas which have increasing abundances of once persecuted species, like the spotted hyena, wild dogs' observed decreases are of up to 30%. The reason was not clear until a few years ago. Then a group of Cambridge scientists (Courchamp et al. 1999) showed that it is exactly the Allee effect that negatively affects the populations of wild dogs. In fact they have a quite peculiar social life: when the juveniles reach the reproductive age, they leave the pack with a group of at most six other individuals of the same sex. A new pack is formed when one of these groups meets a group of the opposite sex. Then a pair becomes dominant and only these two animals reproduce, while the remaining dogs hunt and take care of the puppies. The Cambridge researchers have wondered whether there is a minimum pack size below which survival becomes difficult and found that the threshold exists and consists of three or four adults plus the dominant pair. Below this threshold, the group makes a difficult living, because a good number of "helpers" is necessary for cooperative hunting and defence against attacks by lions and hyenas, especially for the protection of puppies. Also small packs result in small cohorts of juveniles

Fig. 2.4 A group of wild dogs (*Lycaon pictus*)

Fig. 2.5 American ginseng (*Panax quinquefolium*) plants (**a**) and roots (**b**)

available for the colonization of new areas and so the problem is perpetuated along generations.

Even plants can display depensation effects. A documented example (Hackney and McGraw 2001) is that of American ginseng (*Panax quinquefolium*), a perennial plant (see Fig. 2.5) that is grown for its root. The plant also grows in the wild and in this case the root's value is more than tripled. The survival of wild populations critically depends on pollination. It looks like for small populations the rate of pollination is reduced due to the fact that it is more difficult for pollinators to find ginseng plants. As a result the average number of fruits per flower and the average number of fruits per plant is an increasing function of plant density, as shown in Fig. 2.6.

Another possible depensation mechanism is linked to the probability of finding a partner of the opposite sex with whom to mate: for populations that are naturally

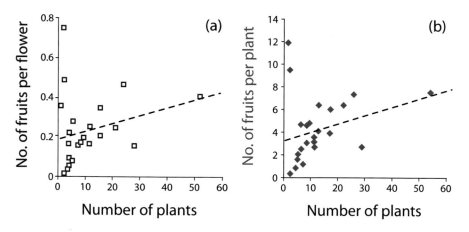

Fig. 2.6 Reproductive success of American ginseng, in terms of number of fruits per flower (**a**, red) and number of fruits per plant (**b**, green) as a function of density (Hackney and McGraw 2001)

low in numbers such as bears and whales, which can anyway launch signals even to great distances, the pairing probability is drastically reduced in very small populations, because the average distance between individuals is so great that signals can be received with much difficulty or are not received. It worthwhile to remark, however, that the population densities at which this effect becomes important are comparable to, if not lower than, those at which demographic stochasticity and genetic deterioration (which will be described later) start operating.

2.4 Simple Dynamic Models of the Allee Effect

It is interesting to explore the consequences of depensation on population dynamics. First, consider a continuous model of density dependence

$$\dot{N} = R(N)N.$$

The presence of an Allee effect implies that the per capita growth rate $R(N)$ is not steadily decreasing with N, rather, for low N values, it is an increasing function of density. For high values of N, it is logical to assume that the phenomenon of intraspecific competition is anyway operating so that $R(N)$ is decreasing with density (see Fig. 2.2 which is a revised version of the original Allee figure). As a result, the per capita rate of demographic increase is a unimodal function of density, as shown in Fig. 2.7a. Sometimes $R(N)$ might even be negative for small N: in this case depensation is termed *critical depensation* (see again Fig. 2.7a), because at low density the mortality rate is larger than the birth rate.

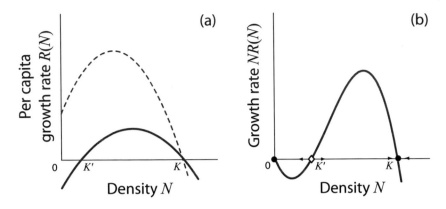

Fig. 2.7 Behaviour of **a** the per capita growth rate $(R(N))$, and **b** the rate of population growth $(NR(N))$ as a function of density N in the case of non-critical depensation (dashed curve) and critical depensation (solid curve)

In the case of *non-critical depensation* there exists only one non vanishing equilibrium population indicated with the label K in the figure. It is easy to understand that this equilibrium population is also stable ($dN/dt > 0$ if $N < K$, while $dN/dt < 0$ if $N > K$) and coincides with the carrying capacity. Therefore the effect of depensation operates only during the transient phase of population growth: if the population were reduced to small numbers, the demographic recovery would be very slow at the beginning and therefore this will increase the probability that the population is drawn into an extinction vortex due to the phenomena we are going to study later (demographic, environmental, genetic stochasticity, etc.), however, if these phenomena did not occur, the population could slowly recover and reach the carrying capacity.

If depensation is critical there exist two non-null equilibrium values, indicated in the figure with the labels K and K'. Because in this case $R(N)N$ has the shape shown as a solid curve in Fig. 2.7b, it turns out that:

$$\frac{dN}{dt} < 0 \text{ if } N < K' \text{ or } N > K$$

$$\frac{dN}{dt} > 0 \text{ if } K' < N < K.$$

Therefore K' is an unstable equilibrium, while K and the null population (extinction) are stable. The important result is thus that the Allee effect can produce a population threshold K' below which extinction is certain.

We already mentioned that one of the causes of the effect can be the increased risk of predation to which small populations are exposed (like in the common guillemot example previously mentioned). It is easy to use the theory of the predator's functional response (Holling 1966) to obtain this intuitive result. Suppose that the population under study does not display inverse density dependence when predators are absent.

We can assume for example that the population has logistic demography. Consider then the case in which predators are present and assume that these predators have a considerable variety of available prey, so their density (which we denote by Y) does not depend on the density of the organisms we are considering. If predators are generalists, such as seagulls that prey on guillemots, we can assume that Y, at least in the area we are studying, is practically constant. Indicate the predator's functional response by $p(N)$ and the per capita growth rate with no predators by $R^*(N)$; then the dynamics of N is given by the equation

$$\dot{N} = NR^*(N) - Yp(N).$$

If the demography is logistic and the functional response is of the second type, the equation becomes

$$\dot{N} = rN\left(1 - \frac{N}{K}\right) - Y\frac{\alpha N}{N + \beta}$$

where r is the intrinsic rate of demographic increase, β the semi-saturation constant and α the maximum rate of consumption of each predator. Therefore the per capita growth rate $R(N)$ that actually governs the population in the area where Y predators are present is given by

$$R(N) = r\left(1 - \frac{N}{K}\right) - Y\frac{\alpha}{N + \beta}.$$

The possible shapes of the population growth rates $NR(N)$ are reported in Fig. 2.8. The shape of $R(N)$ is actually unimodal and therefore gives rise to depensation. It

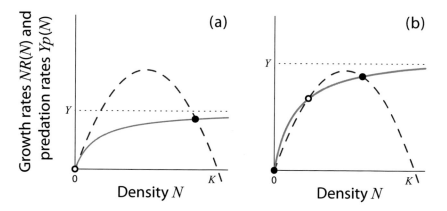

Fig. 2.8 Demographic growth rates of a population in which the functional response of generalist predators can cause depensation. **a** For low predator density, depensation is not critical; **b** for high predator density, depensation is critical. The dashed curve is the population growth rate without predation, the orange curve is the predation rate

is easy to understand that the growth rate is lower for small populations because the mortality rate due to predation is given by

$$\frac{Y\alpha}{N+\beta}$$

and thus the risk of death decreases with N. This depensation is critical if Y exceeds a certain threshold (see panel b in Fig. 2.8). It is easy to compute the threshold by requiring that $R(0)$ be negative, namely that

$$r - Y\frac{\alpha}{\beta} < 0.$$

We obtain the result that depensation is critical for $Y > \beta r/\alpha$, namely when the number of predators is high enough.

2.5　Genetic Deterioration: Basics

The second class of phenomena that can lead to extinction is linked to genetic deterioration. To understand these phenomena it is necessary to introduce some simple concepts of cell biology and genetics, which are reported hereafter.

2.5.1　Cell Structure and DNA

The structural and fundamental unit of living organisms is the cell, which contains the genetic material (the hereditary information), as well as a wide variety of molecules and structures that allows it to maintain its vital functions. The hereditary information is stored in macromolecules, the most important of which is the DNA (deoxyribonucleic acid). The cell is the smallest unit of an organism that is able to function autonomously. Each cell is spatially delimited by an outer membrane, the *cell membrane*, which has the function of separating the content of the cell from the external environment and regulates the flow of substances entering and exiting from the cell thanks to its selective permeability. Outside the membrane, the cells of various organisms (bacteria, plants, fungi) are further bounded by a so-called *cell wall*, which essentially performs a structural function.

Inside the membrane we find the *cytoplasm*, a fluid substance (with high water content) that contains a variety of molecules and specialized structures (called *organelles*). There are two types of cells, mainly distinguished by their different degrees of organization complexity. The cells of *prokaryotes* (from Greek *pro-*, "before", and *karyon*, "nucleus", Fig. 2.9a) are of a smaller size—the diameter is generally comprised between 1 and 5 μm—and have a very simple internal struc-

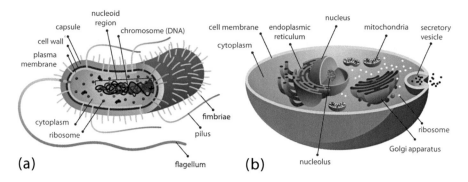

Fig. 2.9 Structure of a prokaryotic cell (**a**) and of a eukaryotic one (**b**)

ture; their genetic material is not separated from the cytoplasm by a membrane, although it predominantly occupies a region of the cytoplasm called *nucleoid*. The cells of *eukaryotic* organisms (from the greek *eu-*,"good", and *karyon*, "nucleus", Fig. 2.9b) have a larger size (10 to 50 μm) and their genetic material is enclosed within a membrane, called the nuclear envelope, which delimits the *nucleus*. The genetic material (the hereditary information) governs the activities of the cell and allows it to transmit its characteristics to offspring.

Biological inheritance, i.e. the process of transmission of individual traits from parents to their offspring, has always been an object of astonishment. However, only in the second half of the nineteenth century, thanks to the studies by Mendel, scientists started to clarify the functioning of hereditary mechanisms. Mendel demonstrated that inherited traits are transmitted as discrete units—called *genes*—which are distributed according to certain rules from one generation to the next. The organisms' genes are located on the chromosomes, complex structures formed by proteins and a macromolecule, the DNA. This nucleic acid, whose structure (see Fig. 2.10) was discovered in the 1950s by Watson and Crick (1953), is like a spiral staircase, or a double-stranded helix. The handrails of the staircase are formed by an alternation of sugar molecules (deoxyribose) and phosphate, while the steps are made of four nitrogenous bases: adenine (A), cytosine (C), guanine (G), thymine (T). Each step consists of two bases and each base is tied to a unit sugar-phosphate. The base pairs are bound together by hydrogen bonds and, because of their structure, adenine can pair only with thymine and cytosine only with guanine. The combination of a base and a unit sugar-phosphate is called nucleotide. DNA is therefore a chain of nucleotides. The sequence of three nucleotides or triplet, in the DNA molecule, represents the genetic code because each triplet codes for the synthesis of a given amino acid (amino acids are the basic components of proteins).

The base sequences contained in the DNA provide the information necessary for protein synthesis which takes place within the cell and is at the basis of life functioning. The DNA can self-replicate, thus allowing the transfer of genetic information from one cell to another. During replication, the chromosomes become clearly visible within the nucleoid (prokaryotes) or the nucleus (eukaryotes). Usually the

Fig. 2.10 The double-stranded helical structure of DNA, proposed by Watson and Crick (1953). The steps of the staircase are made of the four nitrogenous bases that form the nucleotides (counter-clockwise from top left): adenine, cytosine, guanine and thymine. Each step consists of two bases (a base pair)

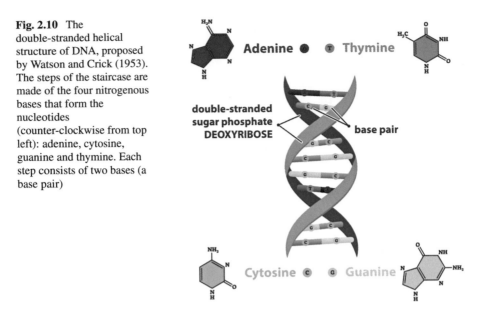

prokaryotes (e.g. bacteria) have a single circular chromosome, while eukaryotes have different chromosomes with a linear structure.

Fundamental to the cell is being able to exchange energy and matter internally and with the external environment. This occurs through various processes of physical and chemical nature, called metabolism. The metabolism is divided into anabolism (which produces complex molecules, useful to the cell, from simpler molecules and requires energy) and catabolism (which involves the degradation of complex molecules into simpler molecules and produces energy). In eukaryotic cells the real engine is constituted by an organelle called mitochondrion. This organelle produces adenosine triphosphate (ATP) from sugars and oxygen. ATP is the high-energy compound required by the vast majority of metabolic reactions. The mitochondrion also contains DNA, which is however different from that of the nucleus.

2.5.2   Cell Cycles and Reproduction

Cell division is the fundamental process that allows a cell replication. In unicellular organisms this process obviously also corresponds to the reproduction of the whole organism. In prokaryotes (e.g. bacteria) reproduction simply occurs via binary fission. In a point of the circular chromosome the double helix begins to separate and the two separate strands act each as a template for a new complementary strand. In this way two chromosomes are formed, each of which migrates to opposite sides of the cell wall. Then the cell stretches itself and separates giving rise to two new sister cells with the same genetic heritage.

More complex is the cell cycle in eukaryotes. It consists of five stages, summarized in Fig. 2.11. The fundamental phase is of course the division of the nucleus, which is termed mitosis. It is divided into four sub-phases: prophase, metaphase, anaphase and telophase, also summarized in Fig. 2.11. In the prophase chromosomes condense and become visible under a microscope. Each of the chromosomes consists of two replicates, named sister chromatids, still joined in a narrow common region, called centromere. Subsequently, the nuclear membrane disintegrates while a spindle-shaped structure forms which consists of microtubule fibres, whose ends are grouped around two pairs of centrioles. Some microtubules connect each pair of centrioles to centromeres. In the metaphase chromatid pairs move back and forth within the spindle until they are arranged symmetrically in the equatorial plane of

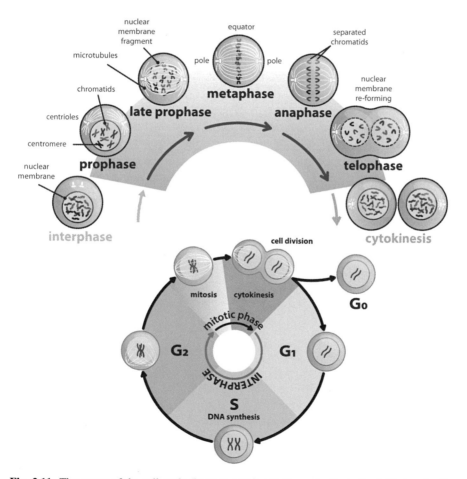

Fig. 2.11 The stages of the cell cycle (bottom panel) and those of mitosis (top). During the G_1 phase there occur cell growth and organelles replication; during the S phase the cell duplicates the chromosomal material; during the G_2 phase are assembled the necessary structures for mitosis and cytokinesis, namely the duplication of the cytoplasm

Fig. 2.12 Karyogram representing the 23 pairs of chromosomes from a female human cell (courtesy of Dr Carola Hartel, GSI Helmoltz Center for Heavy Ion Research, Germany). X are the sex chromosomes

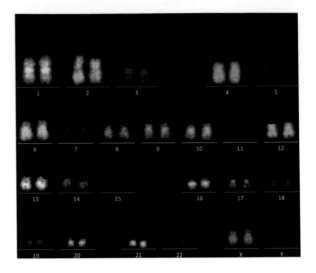

the cell. In the anaphase centromeres separate completely in all pairs of sister chromatids. The chromatids of each pair move far away from each other, so that that each chromatid becomes an independent chromosome attracted to its pair of centrioles. In the telophase the two identical sets of chromosomes reach the two opposite ends of the cell while the spindle fibres degenerate. Subsequently, around each set of chromosomes a nuclear membrane is formed while the chromosomes unwind and again become non visible and the cytoplasm begins to divide.

While binary fission in prokaryotes coincides with reproduction and involves a single circular chromosome, things are more complicated in eukaryotes which possess more chromosomes with a linear structure. In this regard, we have to first specify that chromosomes may or may not be organized in pairs of homologous chromosomes. The two homologues are usually similar in shape and size. When chromosomes are organized in pairs one speaks of diploid cells while in the other case the cells are termed haploid. The clearest example is actually that of a human, whose cells are diploid as they have 46 chromosomes arranged in 23 pairs (see Fig. 2.12). There is an exception for special cells, called gametes (eggs in women and spermatozoa in men), which are haploid, as they have only 23 chromosomes. The gametes are produced from diploid cells, which, through a process called meiosis, double their genetic material and then divide into four daughter cells each containing only one of the homologous chromosomes. There exist, although they are not very frequent, fully haploid eukaryotes. For example amoebae (single-celled protists) have more than 500 small chromosomes which are not arranged in pairs of homologous chromosomes. Reproduction in this case occurs through a process substantially similar to binary fission in prokaryotes. The two daughter amoebae have the same genome, that is they are clones of their mother.

Amoeba is an example of asexual reproduction. In fact, the concept of sex is closely linked to the concepts of haploid and diploid. In the case of humans, for example, the two haploid gametes unite during fertilization to form a single diploid cell (the zygote) which then will develop into an embryo. Among the 23 pairs of human chromosomes there exists a pair of sex chromosomes (conventionally indicated with the letters X and Y). The male sex corresponds to the presence of the XY pair, the female sex to the presence of the XX pair. In the process of sperm formation, male diploid cells, through meiosis, give rise to haploid cells (spermatozoa) half of which contain the X chromosome and half the Y chromosome. All the eggs, instead, since they originate from diploid cells of the type XX, contain the chromosome X. If the zygote arises from the encounter of an egg with a spermatozoon containing the X chromosome, the offspring will be a female, if it originates from a type Y spermatozoon the offspring will be a male. The genetic make-up of children is half inherited from mother and half from father, therefore children are not clones of their parents. However, not necessarily must organisms with diploid cells have sexual reproduction, like in men. In particular, reproduction can occur asexually in the following ways (see Fig. 2.13):

- via *budding*: attached to the mother's body develops a complete individual, which, once it reaches the appropriate size, detaches and becomes independent; an example is that of hydra, a freshwater polyp;
- via *vegetative propagation*: many plants give origin to a clone of the same individual by simple division into two parts, usually genetically identical, or because of simple detachment of body portions;
- via *fragmentation*: one part of the body that is detached regenerates a complete individual, as for example in species of the phyla *Annelida* (ringed worms) or *Platyhelminthes* (flatworms).

Sexes are not always separate. There are hermaphrodites (e.g. snails, see Fig. 2.14d) which have both male and female reproductive organs. Even many plants have hermaphroditic flowers possessing both stamens (male reproductive organs) and pistils (female reproductive organs). If male and female reproductive organs are instead separated the plant bears unisexual flowers or cones. In that case they can be:

- located on the same individual (for example in larch, see Fig. 2.14a): the species is then called *monoecious* (i.e. with one house, *oikos* in Greek);
- carried by two different individuals (e.g. in laurel and in holly, see Fig. 2.14b, c): the species is called *dioecious*.

In hermaphroditic or monoecious species, self-fertilization (i.e. the zygote is formed by the fusion of male and female gametes from the same organism) is very rare. In fact, hermaphroditic animals mate anyway with another partner, while male and female parts of hermaphroditic flowers or male and female flowers of monoecious species generally mature at different times. Therefore, this guarantees the genetic reshuffling since the two halves of the chromosomal make-up come from two different individuals. A special case that is a mixture between sexual and asexual reproduction is parthenogenesis (virginal reproduction). This is quite common for example in rotifers, which are one of the most important components of plankton. Females lay diploid eggs, which do not require fertilization. Therefore, there are only females that produce daughters that are clones of themselves.

Fig. 2.13 Examples of asexual reproduction: **a**, budding in *Hydra spp.* **b**, vegetative reproduction over a leaf of the plant *Kalanchoe delagoensis* **c**, fragmentation in free-living flatworm (*Dugesia japonica*), regenerated as two headed from a trunk fragment exposed to the drug praziquantel (after Nogi et al. 2009)

There actually exist even more complex cases than those described above. In particular, in addition to haploid and diploid cells there exist cells with triplets, quadruplets, etc. of homologous chromosomes. For example, salmon are tetraploid (quadruplets of chromosomes), wheat and kiwi are hexaploid (sextuplets). Moreover, the same species can reproduce either sexually or asexually, depending on the conditions. For example, many species of rotifers are normally parthenogenetic, but under stress they can have sexual cycles: special eggs are produced that are haploid and develop into males which in turn fertilize other haploid eggs thus producing diploid eggs that carry the genetic heritage of both the mother and the father.

2.5.3 Genetic Variability

If the process of DNA replication always occurred without errors there would be no species evolution and there would have been no branching of life into the enormous biodiversity that is before our eyes. The first simple organism that appeared on the earth billions of years ago would have continued to replicate without any change.

Fig. 2.14 **a** Twig of a larch (*Larix decidua*), a monoecious conifer: the male cones, yellow, and the female ones, purple, are carried by the same individual. American holly (*Ilex opaca*) as an example of dioecious species. The flower on (**b**) comes from a male plant and displays flowers with stamens which carry pollen; the flower displaying pistils on (**c**) comes from a female plant. **d** Coupling of snails (*Helix lucorum*), a hermaphroditic gastropod mollusc

The genome instead must be on the one hand very stable—after all, one of the main characteristics of the organisms is that they can make remarkably similar copies of themselves—but on the other hand it must be rather flexible, or capable of development, evolution and branching, in other words diversification. The driver of genome variability is the phenomenon of genetic mutation: during mitosis and meiosis there can occur several errors in DNA replication (see Fig. 2.15) which give rise to mutated cells or to mutant individuals (Fig. 2.16).

However, if mutations, basically due to randomness, were the only cause of diversification, we would not see organized and coherent structures in the terrestrial biosphere. Instead, the organization of life is shaped by the process of natural selection, understood and analysed for the first time by Charles Darwin (1859). In fact, in the first place, because of the genome stability, mutations are anyway very rare (a typical rate of mutation of a pair of DNA bases per generation is in the order of 10^{-6}); also, the vast majority of mutations are deleterious. Among those not detrimental to the survival of organisms only a few manage to pass through the sieve of natural selection, which favours those mutants that have a demographic advantage. Selection is very strict, but the sieve operates continuously on myriads of organisms. It is therefore very effective in the long run and produces organisms adapted to their environment.

In very many cases, as pointed out in more recent times by Kimura and Crow (1964), may nevertheless appear and remain in a population even so-called neutral mutations—i.e. neither advantageous nor deleterious. This makes it possible for the same species and more in particular the same population to be generally characterized by high genetic diversity. Keeping rare genes within a population can not confer an immediate advantage but may be nevertheless useful in the future for the conservation of the species when new environmental conditions might change the selective pressures. The loss of genetic variability can limit the ability of a population to respond to external changes in the long term such as those due to pollution, climate change, new emerging diseases. Unfortunately, in small populations, which is typically the case of populations threatened by extinction, purely random processes can lead to the disappearance of genes thus further increasing the risk of extinction in the long run. In the following sections we will analyse these phenomena in greater detail.

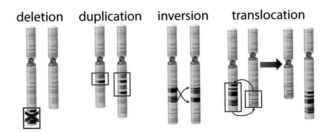

Fig. 2.15 The various types of genetic mutation

Fig. 2.16 The four-winged fruit fly *Drosophila melanogaster*: in the homeotic mutant bithorax, the halteres, which are a balancing organ, are transformed into a second pair of wings (redrawn after Kirschner 2013)

2.6 The Hardy-Weinberg Law

To study the genetic deterioration and its consequences on the process of extinction, we must understand how the genetic structure of a population can change in time. The simplest model that can explain its dynamics is the so-called *Hardy-Weinberg law*. To illustrate and derive it, it is necessary to introduce some basic concepts of population genetics.

As we said in the previous section, errors may occur in the transcription of a gene during DNA replication. The location of the gene along the chromosome is called *locus*. Because of mutations, in a certain locus there can be different variants of a gene, which are called *alleles*. For simplicity, we will assume from now on to have only two alleles denoted by A and a. Suppose we have to deal with organisms with diploid cells—pairs of chromosomes—so that organisms can be, at one locus of the chromosome pair, of type AA or aa or Aa. These variants of an organism of the same species are called *genotypes*. Genotypes AA and aa are called *homozygotes* (organisms derived from homogeneous zygotes) while genotypes Aa are called *heterozygotes* (organisms derived from heterogeneous zygotes). The expression of the gene pool in the visible characters of an organism is called *phenotype*. Genotype and phenotype are not always in one-to-one correspondence because different genotypes may sometimes correspond to the same phenotype. For example, in humans the "eye color" phenotype can be very roughly classified into two categories (brown eyes and blue-green eyes) which are controlled by two alleles (gene A for brown and gene a for blue-green). AA homozygotes and Aa heterozygotes both have brown eyes, while only aa homozygotes have blue-green eyes. In this case we say that the A allele is *dominant* and a is *recessive*. Usually, one indicates the dominant allele with a capital letter and the recessive with a small letter. Not always is there complete dominance: many times the phenotype of the heterozygote is intermediate between the phenotypes of the homozygotes. For example, the coat of a heterozygous mammal with parents one black-coated and the other one white-coated could be grey.

In a population of total size N denote the number of genotypes AA, Aa and aa respectively with N_{AA}, N_{Aa} and N_{aa}. Then we define the genotype frequencies of homozygous dominants, heterozygotes and homozygous recessives as

$$D = \frac{N_{AA}}{N}, \qquad H = \frac{N_{Aa}}{N}, \qquad R = \frac{N_{aa}}{N}.$$

Their sum is obviously equal to one because

$$N = N_{AA} + N_{Aa} + N_{aa}.$$

Within the population of size N there are $2N$ genes of type A and a, since the cells are diploid. We call *gene frequencies* or *allele frequencies* the proportions of each allele in the population. Obviously it turns out that the frequency p of allele A and q of allele a are given by

$$p = \frac{2N_{AA} + N_{Aa}}{2N}, \qquad q = \frac{2N_{aa} + N_{Aa}}{2N}$$

and of course $p + q = 1$.

Therefore the allele frequencies are solely determined by the genotype frequencies because

$$p = D + \frac{H}{2}, \qquad q = R + \frac{H}{2}.$$

Instead, the opposite is not true, that is in general one cannot derive the genotype frequencies from the allelic ones in an unequivocal way. To that end, we have to explicitly introduce some specific assumptions on the demographic and genetic mechanisms that govern the dynamics of the population. The Hardy-Weinberg law is just a relationship between allele frequencies and genotype frequencies which in its simplest formulation is based on the following assumptions. First, suppose we consider a locus with two alleles A and a (the extension to the more complex case of three or more alleles is left to the reader as an exercise). Further assume that

- the population has concentrated reproduction and is semelparous (i.e. none of the adults survive after reproduction);
- the male and female gametes join randomly (for example, females may release eggs and males sperm into the environment, so that eggs and spermatozoa join randomly);
- the genotype frequencies are equal in males and females;
- there is no selective pressure, namely the three genotypes AA, Aa and aa have the same ability to survive and reproduce;
- there is no immigration or emigration, i.e. there is no gene flow between the target population and other populations;

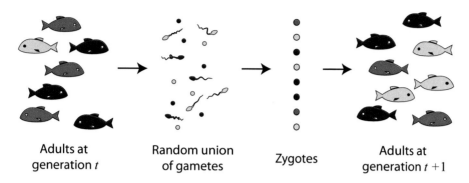

| Adults at generation t | Random union of gametes | Zygotes | Adults at generation $t+1$ |

Fig. 2.17 Life cycle scheme of a semelparous population with random union of gametes

- there are no mutations during reproduction, that is the rate of mutation—which is anyway very small in reality—is supposed *de facto* null;
- the population is very large - theoretically infinite.

Figure 2.17 reports a schematic diagram of the life cycle of a semelparous population with random union of gametes which we will use to derive the Hardy-Weinberg law. Let N_t be the number of adults of the t-th generation and let D_t, H_t and R_t be the genotype frequencies of adults in the t-th generation. Then the allele frequencies in adults are

$$p_t = D_t + \frac{H_t}{2}$$

$$q_t = R_t + \frac{H_t}{2}.$$

Since there is no selection and all genotypes are thus equally fertile, the number of gametes (haploid cells) carrying the gene A is $f(D_t + H_t/2)N_t$ and that of gametes carrying the gene a is $f(R_t + H_t/2)N_t$, where f indicates fertility. It follows that the allele frequencies in the gametes are equal to the allele frequencies that characterized adults. As the population is very large we can identify, using the law of large numbers, frequencies with probabilities and state that, because of the random union of gametes,

- the probability that a zygote AA is formed is p_t^2
- the probability that a zygote Aa (or equivalently aA) is formed is $2p_t q_t$
- the probability that a zygote aa is formed is q_t^2.

Because of the very large number of zygotes (this hypothesis is essential), we can identify these probabilities with the genotype frequencies in zygotes. Finally, since

- there is no selection and thus all zygotes, whatever their genotype, have equal survival up to the adult age,
- adults do not survive from one generation to the next,

we can state that the genotype frequencies in adults at the generation $t + 1$-th are equal to the frequencies in the zygotes. Therefore we conclude that

$$D_{t+1} = p_t^2$$
$$H_{t+1} = 2p_t q_t = 2p_t (1 - p_t)$$
$$R_{t+1} = q_t^2 = (1 - p_t)^2.$$

Frequencies p_t^2, $2p_t q_t$ and q_t^2 are called Hardy-Weinberg frequencies. Note that they are different from the genotype frequencies at the generation t. Instead, the allele frequencies do not change between generation t and $t + 1$, because

$$p_{t+1} = D_{t+1} + \frac{H_{t+1}}{2} = p_t^2 + p_t (1 - p_t) = p_t$$

$$q_{t+1} = R_{t+1} + \frac{H_{t+1}}{2} = (1 - p_t)^2 + p_t (1 - p_t) = 1 - p_t = q_t.$$

Owing to this second result, the genotype frequencies at generation $t + 2$ can be deduced with exactly the same reasoning as above, and therefore—given the time invariance of the allele frequencies—they will be equal to the Hardy-Weinberg frequencies. And so on for the subsequent generations.

We can therefore state the Hardy-Weinberg law saying that, under the hypotheses listed above, the genotype frequencies converge, in a single generation, to the equilibrium values

$$D = p_0^2 \qquad H = 2p_0 q_0 \qquad R = q_0^2$$

where p_0 and q_0 indicate the initial allele frequencies.

The Hardy-Weinberg law is fairly robust: it holds in fact under less restrictive assumptions than those listed above (although the derivation of the law becomes harder). In particular, it is not necessary to assume discrete generations or equal genotype frequencies in males and females; moreover, instead of random union of gametes released in the environment, one can assume random mating of males and females and internal fertilization. With these less restrictive assumptions, there is still convergence towards the Hardy-Weinberg frequencies, although convergence can take longer than a generation. It is important to remark that if a gene is present with very low frequency in the initial population, the Hardy-Weinberg law stipulates that this gene will remain indefinitely in this population and therefore heterozygous individuals will indefinitely be present in the population, thus transmitting both genes to the next generation. However, this conclusion is crucially based on the assumption that the population is very large, which is not at all verified in the case of populations that are threatened by extinction. To analyse the problem of the extinction risk, one must then go beyond the findings summarized in the Hardy-Weinberg law.

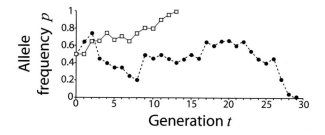

Fig. 2.18 The effect of genetic drift in a small population consisting of 10 individuals. The two random simulations show that there can occur both the loss of gene *A* with the consequent fixation of gene *a* (black circles, after 30 generations) and the fixation of *A* with the consequent loss of *a* (white squares, after 13 generations). The two genes are initially present with equal frequency

2.7   Genetic Drift and Wright's Formula

When studying small populations it is essential to introduce the phenomenon of *genetic drift*. It was Sewall Wright (1931) who pointed out that in a finite population there is a decreasing trend of heterozygosity *H* from one generation to the next. In particular, there can be loss of one allele and fixation of the other even if alleles are mutually neutral, that is even if one gene does not provide any demographic advantage over the other and thus natural selection is not operating. Figure 2.18 shows, through random simulations, that the fate of two genes undergoing no selection and initially present with equal frequencies ($p_0 = \frac{1}{2}$) in a population of 10 individuals can be completely different depending on the sequence of random events. Note that, after a sufficient number of generations, the heterozygosity of the population is reduced to zero. The Hardy-Weinberg law, instead, would predict that after a sufficient number of generations the heterozygosity would necessarily converge to $2p_0(1 - p_0) = \frac{1}{2}$.

To analyse the phenomenon of genetic drift from a theoretical viewpoint, we again consider a population that meets all the assumptions leading to the Hardy-Weinberg law except the one about the size of the adult population, i.e. it is no longer considered to be very large. It is easier to derive how heterozygosity varies in zygotes rather than in adults by using a simplified scheme like the one described in Fig. 2.19. In agreement with what occurs very frequently in nature, suppose that a finite number of adults indicated by *N* produces nevertheless a large number of gametes (phase 1 in the figure) which join randomly forming an equally large number of zygotes (phase 2). However, for reasons related to resource availability and/or the characteristics of the territory in question, it is realistic to assume that at generation *t* only a finite number N_t of individuals survives (phase 3 in the figure).

Suppose that in the large pool of gametes that eventually will give rise to the adults at generation *t*, there is a fraction p_t of type *A* alleles. The gametes join two by two, to form the zygotes which will then become adult. Since the union of gametes is random and both gametes and zygotes are in large number the Hardy-Weinberg law applies and one can deduce that the allele frequency in the zygotes will again be p_t, while

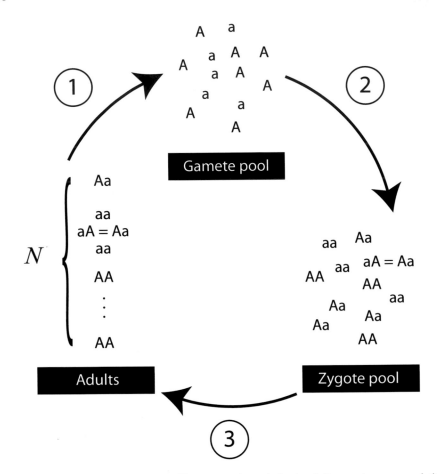

Fig. 2.19 Scheme for the calculation of heterozygosity variation in a finite semelparous population

the frequency of heterozygotes will be $H_t = 2p_t(1 - p_t)$. As previously stated, only a finite number N_t of zygotes, from among the large number of those generated, will survive to adulthood. Since no natural selection is operating according to our assumptions, different genotypes have equal survival probability. We wonder which number j of type A genes will be present in the adults. As the process of survival is random, this is equivalent to tossing up an allele $2N_t$ times with the allele A having a probability p_t to be chosen, because p_t is the frequency of the gene in the pool of zygotes. Therefore the probability distribution $P(j)$ of j is a binomial, namely

$$P(j) = \binom{2N_t}{j} p_t^j (1 - p_t)^{2N_t - j} = \frac{(2N_t)!}{j!(2N_t - j)!} p_t^j (1 - p_t)^{2N_t - j}.$$

Recall that the average value (indicated hereafter with E[·]), the variance (indicated with Var[·]) and the second moment of a binomial random variable are given by

$$E[j] = 2N_t p_t$$
$$Var[j] = 2N_t p_t (1 - p_t)$$
$$E[j^2] = Var[j] + (E[j])^2 = 2N_t p_t (1 - p_t) + 4N_t^2 p_t^2.$$

In particular, we can also calculate the mean, the variance and the second moment of the allele frequency in adults, that is the random variable $\frac{j}{2N_t}$. Of course, it turns out

$$E\left[\frac{j}{2N_t}\right] = \frac{E[j]}{2N_t} = p_t \tag{2.1}$$

$$Var\left[\frac{j}{2N_t}\right] = \frac{Var[j]}{4N_t^2} = \frac{1}{2N_t} p_t (1 - p_t) \tag{2.2}$$

$$E\left[\left(\frac{j}{2N_t}\right)^2\right] = \frac{E[j^2]}{4N_t^2} = \frac{Var[j] + (E[j])^2}{4N_t^2} = \frac{1}{2N_t} p_t (1 - p_t) + p_t^2. \tag{2.3}$$

When adults reproduce, the allele frequency in the gametes, which are large in numbers, will still be $\frac{j}{2N_t}$. On the other hand, when the gametes of generation $t + 1$ randomly join to form a large number of zygotes, the frequency of heterozygotes H_{t+1} will follow the Hardy-Weinberg law, that is

$$H_{t+1} = 2\frac{j}{2N_t}\left(1 - \frac{j}{2N_t}\right).$$

Since j is a random variable, the frequency of heterozygotes is random too. We can compute the expected frequency of heterozygotes at generation $t + 1$, using the relationships (2.1)–(2.3):

$$E[H_{t+1}] = \sum_{j=0}^{2N_t} 2\frac{j}{2N_t}\left(1 - \frac{j}{2N_t}\right) P(j) =$$
$$= 2\left\{E\left[\frac{j}{2N_t}\right] - E\left[\left(\frac{j}{2N_t}\right)^2\right]\right\} =$$
$$= 2\left(p_t - \frac{1}{2N_t} p_t (1 - p_t) - p_t^2\right) = 2p_t (1 - p_t)\left(1 - \frac{1}{2N_t}\right).$$

As $H_t = 2p_t (1 - p_t)$, then the resulting Wright's formula for genetic drift is

$$E[H_{t+1}] = H_t\left(1 - \frac{1}{2N_t}\right).$$

In other words, the average value of the frequency of heterozygotes decreases from one generation to the next by a factor $1 - \frac{1}{2N_t}$. Quite evidently, the smaller the population the more important this factor is. A very similar formula can be obtained considering the frequency of heterozygotes in the adult population. More precisely, since the heterozygosity of the adults at generation t is a random variable too, it turns out

$$\mathrm{E}\left[H_{t+1}\right] = \mathrm{E}\left[H_t\right]\left(1 - \frac{1}{2N_t}\right).\tag{2.4}$$

If the population is not made up of hermaphrodites or by an equal number of males and females, all reproductive, Wright's formula still holds provided the total number of individuals is replaced by the so-called *effective population size* $N_{e,t}$, which is the number of individuals that actually breed in the t-th breeding season. In very many populations, $N_{e,t}$ may be only a small fraction of the total population size N_t. If we denote by $\overline{H}_t$ the average heterozygosity at time t the following formula holds

$$\overline{H}_{t+1} = \overline{H}_t\left(1 - \frac{1}{2N_{e,t}}\right).$$

Often, populations are considered that are at demographic equilibrium, namely in which $N_{e,t} = N_e = $ constant. In this case it turns out

$$\overline{H}_t = \overline{H}_0\left(1 - \frac{1}{2N_e}\right)^t.$$

Figure 2.20 graphically shows that in populations with effective size of 5–10 individuals the heterozygosity loss over 10 generations is between 40 and 60%. Populations with effective size higher than 50–100 individuals lose less than 1% of their heterozygosity from one generation to the next. After 10 generations, a population of 50 individuals still keeps 90% of its initial heterozygosity.

It is obvious, but important, to note that the level of heterozygosity in a population is anyway linked to the initial heterozygosity $\overline{H}_0$. This simple remark is the essence of the so-called *founder effect*. It occurs whenever a few individuals leave a large population to create a new one, by immigrating to a new suitable habitat. These founders, by chance, may not be representative of the genetic variability of the whole species. The initial heterozygosity can thus be very low and genetic drift further depresses the genetic variability in the new population. One famous case is that of the lion population inhabiting the Ngorongoro crater in Tanzania (Primack 2000). In 1962, the original population was decimated by an explosion of blood-sucking insects and reduced to 9 females and 1 male. Two years later another 7 males immigrated bringing the number of founders of the new population to 17. Compared to the nearby much larger lion population in the Serengeti, Ngorongoro's lions display a very low genetic variability which is reflected in their difficulty to grow demographically: the population after reaching a size of more than 100 individuals was then reduced to about 40 individuals in 2000.

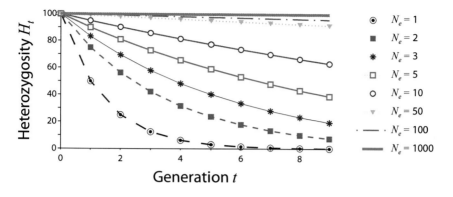

Fig. 2.20 Loss of heterozygosity (ratio between the percentage H_t of heterozygotes in the tth generation, and the initial percentage H_0 of heterozygotes) versus time as a function of different effective population sizes N_e, assumed to be constant over time

One of the causes of reduced demographic growth in small populations with little genetic variability is the so-called *inbreeding*, namely the phenomenon of mating between related individuals (mother with son, uncle with niece, self-pollination in hermaphrodite flowers). We previously stated that there exists biological mechanisms that hinder inbreeding, such as blooming of male and female flowers at different times or the emigration of young males far away from their family. In small populations, though, these mechanisms are no longer so effective, for example due to the lack of reproductive partners that are not consanguineous. Therefore, there is a higher likelihood for the fixation of recessive deleterious alleles, which make offspring little viable or infertile. Considering again the example of Ngorongoro's lions, it is worth remarking that the population experienced, in addition to the founder effect, the phenomenon of inbreeding: as a result sperm abnormalities were observed in many males (Fig. 2.21) with a consequent decrease in fertility.

The effective population size is crucially influenced by the sex ratio and the mating mode. We denote by N_m and N_f the number of males and females in the population, so that $N_m + N_f = N$. If we are considering a monogamous species (like e.g. many birds) with a different number of males and females, evidently the number of pairs that are formed is determined by the sex represented by the smaller number of individuals, or

$$N_e = 2 \min(N_m, N_f).$$

If instead we are considering a polygamous species, there can be different situations: sometimes a male monopolises most females (polygyny, like e.g. in the elephant seal and many marine mammals of big size), while more rarely a female monopolises most males (polyandry, for example in some monkeys). In many cases, on the contrary, it is reasonable to assume that mating is random: a good model is the one in which the probability for a female (respectively for a male) to mate increases linearly with the

percentage of males (respectively females) in the population and is equal to 1 when
the sex ratio is 1:1. It follows that

$$N_e = N_f \frac{2N_m}{N_m + N_f} + N_m \frac{2N_f}{N_m + N_f} = \frac{4N_f N_m}{N_m + N_f}.$$

The loss of genetic diversity due to random drift can be countered by mutations and
gene flow, phenomena that we assumed to be absent in our simplifying assumptions.
The mutation rate is, however, very small, as we have seen. Quantitative analyses, not
reported here, show that in populations with fewer than 100 individuals mutations
cannot at all counteract genetic drift in a significant way. Much more effective is the
phenomenon of gene flow (due to migration between different populations of the
same species). Just one new individual per generation immigrating into a population
is sufficient to effectively counter genetic drift. Immigration of 5–10 individuals per
generation make the effect of drift negligible. One should not forget, however, that
the reduction in the number of individuals in a population is often accompanied by
the fragmentation and insularisation of habitats (see the later chapter on metapopulations). Populations become not only smaller but also more and more isolated, which
greatly reduces the likelihood of gene flow.

Fig. 2.21 Pleiomorphic sperm forms in the ejaculates of Ngorongoro lions. Various abnormalities,
including macrocephalic (**e**) and microcephalic (**f**) spermatozoa, are shown for comparison with a
normal spermatozoon (**a**) (redrawn from Wildt et al. 1987)

Often, the size of many populations fluctuate over time for reasons unrelated to the genetic deterioration but due to exogenous causes, such as changes in resources availability or climatic conditions. In such a case the trend of heterozygosity in time is given by

$$\overline{H}_t = \overline{H}_0 \left(1 - \frac{1}{2N_{e,0}}\right)\left(1 - \frac{1}{2N_{e,1}}\right)\cdots\left(1 - \frac{1}{2N_{e,t-1}}\right).$$

Therefore, if for some reason the effective population size drops to a very small value in a given year, the corresponding Wright's factor is particularly small and the frequency of heterozygotes is drastically reduced, thus affecting the heterozygosity in all subsequent generations. This phenomenon is called a *bottleneck*.

2.8 Effects of Genetic Drift on Population Dynamics

So far we have assumed that the population has a constant size—equivalent to assuming a demography with a finite rate of demographic growth equal to one—or that the population fluctuations are linked to exogenous factors, not intrinsic to the population. We can, however, wonder what may happen if the demographics of the population is also linked to intrinsic factors, such as its density. Many data confirm that the demographic rates depend on both the population size and its genetic structure. For example, in the plant *Ipomopsis aggregata*, typical of Arizona's mountain areas, the success of seed germination significantly depends on the population size (Fig. 2.22). In fact, smaller populations are characterized by smaller plants which produce smaller seeds that are more subject to environmental stress. However, when the plants are artificially pollinated with pollen from large populations (thus characterized by greater genetic variability) the germination success increases significantly. This proves that the rate of demographic growth of the plant is positively influenced by the genetic diversity of each population.

A very simple way to introduce genetic deterioration into a population demographics is to assume that the finite rate of growth is an increasing function of Wright's factor, or in other words the rate of population growth from one generation to the next is lower than optimal if there is a reduction of heterozygosity between generations. For the sake of simplicity we can assume that the finite growth rate is proportional to the Wright factor and equal to the optimal value $\Lambda(N_t)$ when there is no loss of heterozygosity (i.e., when Wright's factor is equal to 1). If we denote by γ the ratio of effective population size to total population size ($N_e = \gamma N$), we can then write the following simple model of population dynamics

$$N_{t+1} = \left(1 - \frac{1}{2\gamma N_t}\right)\Lambda(N_t)N_t. \tag{2.5}$$

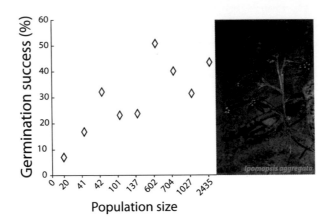

Fig. 2.22 Percentage of germinating seeds in populations of different size for the plant *Ipomopsis aggregata* in Arizona. Reworked from Heschel and Paige (1995)

Fig. 2.23 Moran diagram of a Beverton-Holt model ($\lambda = 3$, $\alpha = 0.06$, curve in grey), modified via the Wright factor (Eq. 2.5). The ratio γ of effective population size to total size is equal to 20%. The non-trivial unstable equilibrium is denoted by U, while the non-trivial stable equilibrium is labelled S

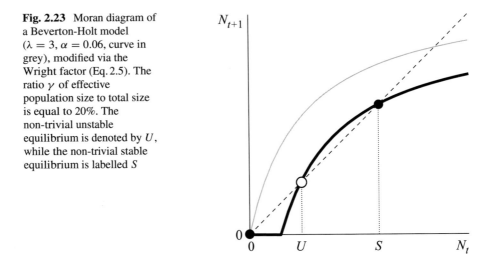

Of course for $\gamma N < \frac{1}{2}$ Wright's factor would be negative and must therefore be set equal to zero. Actually, the effective population size cannot be smaller than 2, because otherwise one of the two sexes would be absent. In other words genetic deterioration in small populations has such a large effect that the rate of population growth vanishes. It is easy to understand that population dynamics as described by Eq. 2.5 is characterized by critical depensation. For example, Fig. 2.23 displays a Moran diagram for a Beverton-Holt model in which the Wright factor has been introduced.

It is worth remarking that in this case extinction is a stable equilibrium for the population. There also exist two non-trivial equilibria: one with lower abundance is unstable and acts as an extinction threshold, while the one with higher abundance is stable. It is therefore apparent that genetic deterioration is a phenomenon that leads to consequences very similar to those we have identified and discussed in Sect. 2.4 of this chapter devoted to the Allee effect.

References

Allee W (1931) *Animal Aggregation. A Study in General Sociology.* University of Chicago Press, Chicago, USA

Beverton R, Holt S (1957) On the Dynamics of Exploited Fish Populations. H.M., Stationery Office, London

Birkhead T (1977) The effect of habitat and density on breeding success in the common guillemot (Uria aalge). Journal of Animal Ecology 46:751–764

Courchamp F, Clutton-Brock T, Grenfell B (1999) Inverse density dependence and the Allee effect. Trends in Ecology and Evolution 14:405–410

Darwin C (1859) On the Origin of Species. John Murray, London, U.K

Hackney EE, McGraw JB (2001) Experimental demonstration of an Allee effect in American Ginseng. Conservation Biology 15:129–136

Heschel MS, Paige KN (1995) Inbreeding depression, environmental stress, and population size variation in scarlet gilia (Ipomopsis aggregata). Conservation Biology 9:126–133

Holling C (1966) The functional response of invertebrate predators to prey density. Memoirs of the Entomological Society of Canada 98:5–86

Iannelli M, Pugliese A (2014) An Introduction to Mathematical Population Dynamics. Springer, Switzerland

Kimura M, Crow IF (1964) The number of alleles that can be maintained in a finite population. Genetics 49:725–738

Kirschner M (2013) Beyond Darwin: evolvability and the generation of novelty. BMC Biology 11:110

Nogi T, Zhang D, Chan JD, Marchant JS (2009) A novel biological activity of praziquantel requiring voltage-operated Ca^{2+} channel β subunits: Subversion of flatworm regenerative polarity. PLOS Neglected Tropical Diseases 3:1–13

Primack R (2000) A Primer of Conservation Biology. Sinauer Associates, Sunderland, MA, USA

Ricker W (1954) Stock and recruitment. Journal of the Fisheries Research Board of Canada 11:559–623

Watson JD, Crick FHC (1953) Molecular structure of nucleic acids. Nature 171:737–738

Wildt D, Busk M, Goodrowe K, Packer C, Pusey A, Brown J, Joslin P, Obrien S (1987) Reproductive and genetic consequences of founding isolated lion populations. Nature 329(6137):328–331

Wright S (1931) Evolution in Mendelian populations. Genetics 16:97–159

Chapter 3
Extinction Risk Analysis: Demographic and Environmental Stochasticity

3.1 Population Fluctuations

The numerical strength of plant and animal populations often displays a more or less large variability which contrasts with the simple deterministic models of population dynamics studied in basic courses of ecology. The cause of this variability is in many cases unknown, thus fluctuations are classified as random or, technically speaking, *stochastic*. It is anyway possible to distinguish between different types of stochasticity and provide a description for each type by using appropriate quantitative models that are not deterministic. These models are very useful in conservation ecology because among other things they allow predictions on the probability of a population future trends and hence the assessment of its risk of extinction.

Figures 3.1, 3.2 and 3.3 contain several examples of the fluctuations in numbers of animal populations (in these specific cases, birds and mammals). For each species, the corresponding graph shows the evolution of the numbers of individuals in different sites. It is important to compare the characteristics of species with short versus long life expectancy. Consider, for example, the three populations of two birds of the genus *Parus* (*Parus caeruleus* and *Parus major*) compared with the mute swan *Cygnus olor*. The blue tit and great tit have lifetimes of about 2 years and age at first reproduction of one year, while the swan has a lifespan of about 10 years and an age at first reproduction of about 4 years. The numbers of blue tits (panel a of Fig. 3.1) display fluctuations much faster and more pronounced than those of the swan (panel b), whose population has a combination of long-term (about 30 years) and short-term (a few years) fluctuations. In fact, populations with a larger number of age classes are able to better absorb variations of the external conditions (for example, food availability and weather vagaries) because often it is only the younger classes that actually suffer from this variability.

As the size N of a population is always a non-negative variable, its variability over time can be effectively described by the coefficient of variation (CV), namely the ratio of the standard deviation of the series of abundances σ_N to the average size of the population $\bar{N}$:

Fig. 3.1 Fluctuations in population abundances of **a** the tit (*Parus spp.*), and **b** the mute swan (*Cygnus olor*). For tits, the population data from Ghent (Belgium) are reported on the left axis, while the population data from Oxford (England) are reported on the right axis. The swan population is also English, living in the River Thames. Data from Lande et al. (2003)

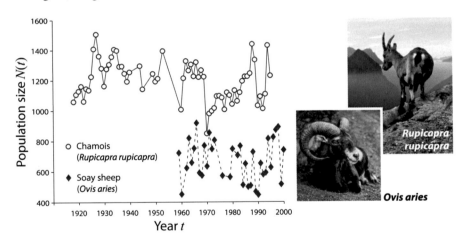

Fig. 3.2 Fluctuations in the abundances of a chamois (*Rupicapra rupicapra*) population in the Swiss National Park and of a Soay sheep (*Ovis aries*) population in the island of Hirta, Scotland. Data from Lande et al. (2003)

$$CV = \frac{\sigma_N}{\overline{N}}.$$

An alternative measure of variability is the standard deviation of the logarithm of size N, i.e. $\sigma_{\log N}$. For CV values lower than 30% the two variability indices are substantially equivalent. Table 3.1 reports the estimated coefficients of variation of

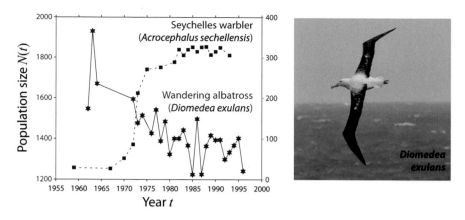

Fig. 3.3 Fluctuations in the abundances of a Seychelles warbler (*Acrocephalus sechellensis*) population at Cousin Island, Seychelles, and of a wandering albatross (*Diomedea exulans*) population in southern Georgia, USA. Data from Lande et al. (2003)

different bird populations. One can note that CV tends to decrease with the increase of the age at first reproduction, thus more generally confirming what we already stated for the fluctuations of tits and the mute swan.

Not always can the differences between average life expectations be sufficient to explain the different characteristics of population fluctuations. Consider, in this regard, the cases of chamois and Soay sheep (or primitive mouflon) shown in Fig. 3.2. While the chamois displays fluctuations with periodicity of about a decade, the Soay sheep has faster fluctuations with periodicity between two and three years. However, the two species have life expectations at birth that are roughly equal (10–12 years). In mammals, then, other factors seem to play a role in determining the stochasticity of population dynamics.

3.2   The Different Types of Stochasticity

The dynamics of each population is a mix of deterministic components (i.e. those that we can understand, predict and measure) and stochastic components (i.e. those that we are not able to understand and/or predict and/or measure). Sometimes, the deterministic components of the dynamics prevail, such as in the Seychelles warbler and the wandering albatross (see Fig. 3.3). The first species, as a consequence of an active policy of conservation, has grown in a roughly logistic way stabilizing at the carrying capacity around which it fluctuates with a small stochastic component. The second species, instead, has suffered a high adult mortality (resulting in negative growth rate) because of poisoned baits attached to long-lines of commercial fishing boats. The consequence has been an exponential decrease of its numbers around which there have been small stochastic variations.

Table 3.1 Coefficient of variation (CV) of the fluctuations in numbers of different bird populations. The CV is computed from the last 20 years only. Parameter α indicates the age at first reproduction of mothers. Data from Lande et al. (2003)

Species	α	Site	Period	CV
Blue Tit *Parus caeruleus*	1	Lower Saxony Germany	1974–1993	0.27
Dipper *Cinclus cinclus*	1	Lygnavassdraget Southern Norway	1978–1997	0.46
Garganey *Anas querquedula*	1	Engure Marsh Latvia	1974–1993	0.48
Great Tit *Parus major*	1	Lower Saxony Germany	1974–1993	0.17
Great Tit *Parus major*	1	Wytham Wood Oxford, England	1974–1993	0.27
Northern Shoveler *Anas clypeata*	1	Engure Marsh Latvia	1974–1993	0.38
Nuthatch *Sitta europaea*	1	Lower Saxony Germany	1974–1993	0.31
Pied Flycatcher *Ficedula hypoleuca*	1	Lower Saxony Germany	1974–1993	0.22
Pied Flycatcher *Ficedula hypoleuca*	1	Lingen Germany	1974–1993	0.25
Pied Flycatcher *Ficedula hypoleuca*	1	Kilpisjärvi Northern Finland	1968–1987	0.45
Pochard *Aythya ferina*	1	Engure Marsh Latvia	1974–1993	0.51
Seychelles Warbler *Acrocephalus sechellensis*	1	Cousin Island Seychelles	1973–1993	0.10
Song Sparrow *Melospiza melodia*	1	Mand arte Island British Columbia, Canada	1979–1998	0.49
Tufted Duck *Aythya fuligula*	1	Engure Marsh Latvia	1974–1993	0.25
Avocet *Recurvirostra avosetta*	2	Havergate Suffolk, England	1967–1986	0.30
Grey Heron *Ardea cinerea*	2	Southern England	1979–1998	0.09
Kentish Plover *Charadrius alexandrinus*	2	Niedersachsen Germany	1974–1993	0.57
Ural Owl *Strix uralensis*	2	Hämeenlinna Finland	1969–1988	0.5
Common Tern *Sterna hirundo*	3	Mecklenburg-Vorpommern Germany	1978–1997	0.39
Sandwich Tern *Sterna sandvicensis*	3	Schleswig-Holstein Germany	1974–1993	0.44
Mute Swan *Cygnus olor*	4	River Thames England	1920–1939	0.15
South Polar Skua *Catharacta maccormicki*	6	Pointe Géologie Terre Adélie, Antarctica	1981- 2000	0.19
Short-tailed Shearwater *Puffinus tenuirostris*	7	Fisher Island Tasmania	1965–1984	0.15
Northern Fulmar *Fulmarus glacialis*	9	Eynhallow Orkney	1958–1977	0.27
Wandering Albatross *Diomedea exulans*	10	Bird Island South Georgia	1974–1993	0.06

Deterministic factors can sometimes cause fluctuations and irregular oscillations that at first glance may seem completely stochastic. This is the so-called phenomenon of *deterministic chaos* which is linked to a strong density dependence (overcompensation), like the one that characterizes for example the Ricker model. However, in populations with no age structure, such as those of univoltine insects, rarely are intrinsic rates of demographic increase so large in reality as to cause large chaotic fluctuations. In populations with a rather long life expectation, hence with several age classes, there can be chaotic fluctuations for lower intrinsic growth rates, which in most cases are, nonetheless, considerably higher than the real ones. When irregular fluctuations of plant and animal population numbers are observed, it is fairly certain that in most cases these are linked with truly stochastic factors and more rarely with deterministic chaos.

As for the mechanisms generating these random fluctuations, we can distinguish among three main types of stochasticity (Lande et al. 2003). The first source of observed population vagaries is, so to say, trivial and will not be explicitly considered in the following: it is the *random measurement error*. Censuses, counts or estimates of a population size (for example, by means of capture-mark-recapture methods) are never perfect and therefore fluctuations of numbers can simply be caused by these sampling errors. The second type of source is the so-called *demographic stochasticity* which depends on random events operating at the level of one single individual mortality and reproduction. Like genetic drift, it acts with particular strength in small populations. The third type is *environmental stochasticity* which depends on random events operating at the level of the environment in which populations live. This kind of stochasticity operates effectively and in a comparable way in both small and large populations.

To better understand what we mean by demographic stochasticity, it should be noted that, even in populations with no age or size structure, individuals are all equal only in the average. In particular, with regard to mortality, in a given amount of time each individual can either die or survive. If then in year t a population is for example composed of 5 individuals, in year $t + 1$ there may be 0 or 1 or 2 or 3 or 4 or 5 surviving individuals. At the individual level, the mortality rate must be treated as the probability that an individual dies in the time unit. Therefore if, in the above example, the population had a survival rate from year to year amounting to 40%, there would be a probability of $0.4^5 \simeq 0.01$ that all 5 individuals survive until year $t + 1$ and a probability of $0.6^5 \simeq 0.077$ that all die. As one can remark, the extinction probability of the population, even in the course of a single year, is not at all negligible. A similar reasoning can take place for reproduction, because every sexually mature individual can only produce an integer number of offspring (0 or 1 or 2, etc.). For example, when we say that the population has a birth rate of 2.5 daughters per mother per year, this value is the average resulting from a probability distribution where some mothers produce 0 daughters, some one daughter, some two daughters, some three daughters and so on. In large populations, it is reasonable to use the law of large numbers and interpret probabilities as replaceable by proportions. If in the example under consideration the population consisted in year t not of 5 but of 1000 individuals, the chances that all survive or die would be infinitesimal and we would

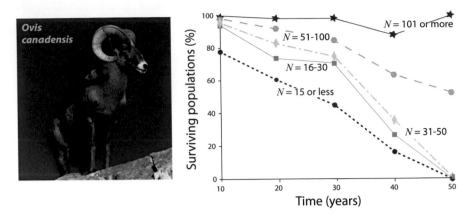

Fig. 3.4 Relationship between the initial size N and the percentage of surviving populations through time for the bighorn sheep (*Ovis canadensis*). Data from 120 populations in south-western US (Berger 1992)

not commit big mistakes saying that in year $t + 1$ there will be approximately 400 surviving individuals. Similarly, we would not commit big mistakes by saying that the newborns (assuming a sex ratio 1: 1) will be approximately 2,500. However, in small populations demographic stochasticity operates quite effectively, so that even with rather high rates of survival and reproduction the extinction risk cannot be overlooked. Figure 3.4 illustrates the effect of demographic stochasticity on different populations of the bighorn sheep (*Ovis canadensis*). Only the populations that initially have more than 100 individuals have survived for longer than 50 years. Figure 3.5 shows the actual decline and final extinction of small populations from four different species. One can note that usually the decline is not sudden, rather relatively gradual.

Unlike demographic stochasticity, environmental stochasticity does not stem from differences between two individuals of the same population, but from temporal variations of the surrounding environment which affect the demographic parameters of all individuals of the population. It is therefore due to seasonal changes of factors external to the population, such as weather (temperature, precipitation, etc.), food availability, the presence of diseases, predation by other species, and so on. For this reason, as already stated, environmental stochasticity does produce effects on both large and small populations, because it influences more or less in the same way the survival and reproduction abilities of each of the organisms that belong to the population. Figure 3.6 displays an example of this type of stochasticity. The reproductive success of flamingos of a South African park is correlated, albeit weakly, with rainfall because years with low precipitation are characterized by more frequent reproduction failures, while years with larger precipitation generally allow a good number of chicks to reach fledging.

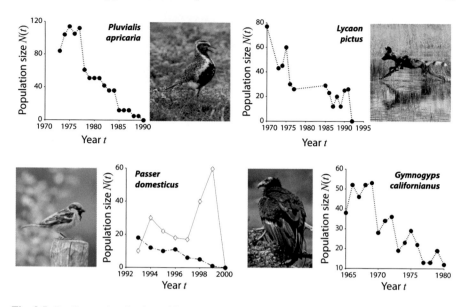

Fig. 3.5 Decline and extinction of four small populations: the golden plover (*Pluvialis apricaria*) in Scotland (Parr 1992), the African wild dog (*Lycaon pictus*) in Serengeti Park, Tanzania (Burrows et al. 1995), two populations of sparrow (*Passer domesticus*) in Helgeland, Norway (Lande et al. 2003), and the California condor (*Gymnogyps californianus*, Dennis et al. 1991)

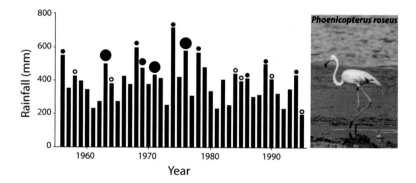

Fig. 3.6 The influence of rainfall on reproductive success in populations of two species of flamingos in the Etosha National Park, South Africa (Simmons et al. 1996). Vertical bars indicate the total annual rainfall recorded in the park. Reproductive success is indicated by circles: open circles indicate reproduction failure (eggs are laid but no chick will be able to fledge), filled circles indicate reproductive events of small, medium or large success depending on the circle size

3.3 Individual Fitness and Models of Demographic and Environmental Stochasticity

We will describe simple cases only, neglecting age and size structure. Usually, we will refer to populations with discrete dynamics in which reproduction is annual and synchronized. We assume that populations either are semelparous (annual plants, univoltine insects) or have overlapping generations, but consist of organisms that start breeding one year after birth and in which the survival of adults (i.e. individuals with age ≥ 1 year) and the fertility are independent of age. For the sake of simplicity, we assume that, if the population reproduces sexually, the sex ratio is constant, so that it is sufficient to consider the dynamics of females only. We will use as the baseline model for our consideration the following

$$N_{t+1} = \lambda_t N_t$$

where N_t represents the number of adult females in year t and λ_t is the finite rate of demographic growth during year t.

Note that the finite rate of population growth can vary from year to year, which is why this rate is indicated as λ_t instead of λ, a constant. The drivers of these variations may be different: density dependence (i.e. the fact that the rate may vary with N_t, namely $\lambda_t = \Lambda(N_t)$), demographic and/or environmental stochasticity, interactions with other populations, etc. Remember that for *semelparous populations* the finite growth rate is

$$\lambda_t = \sigma_{Y,t} f_t$$

where $\sigma_{Y,t}$ is the survival from juvenile to reproductive adult, while f_t is fertility. As for populations with overlapping generations the growth rate is instead given by

$$\lambda_t = \sigma_{Y,t} f_t + \sigma_{A,t}$$

with $\sigma_{A,t}$ indicating the adult survival from year to year.

To discuss stochasticity from a quantitative viewpoint, one must first consider that the rate of demographic growth λ_t is actually the average of the individual contributions of each adult female to the reproductive output in year t and to the survival from year t to year $t + 1$. The contribution of the i-th female to the change of population abundance is termed *individual fitness*. Rather than resorting to convoluted theoretical definitions, it is more appropriate to understand the concept of individual fitness through a simplistic example. One of many females, say the i-th, belonging to a population of e.g. great tits, produces 5 eggs in year t. Four of these eggs hatch, but only 2 of the 4 birds are female and only one of them survives until the first birthday. Also the i-th female succeeds in surviving until the $t + 1$-th breeding season. Therefore, it turns out that

$$\text{fitness}(i, t) = 1 + 1 = 2.$$

If we denote by $w_{i,t}$ the fitness of the i-th female in year t, we can then write that

$$N_{t+1} = \sum_{i=1}^{N_t} w_{i,t},$$

which implies that

$$\lambda_t = \frac{1}{N_t} \sum_{i=1}^{N_t} w_{i,t}$$

is nothing but the average fitness of the t-th season. There is therefore a strong link between the finite rate of population growth and individual fitnesses. It is also instructive to break down each individual fitness into the sum of two terms. The first term is the expected value of the fitness of the i-th female (say $\overline{w}$), while the second term Δ is the deviation from the mean. It is almost always reasonable to assume that, while the expected value of the fitness depends on season t, the deviation from the average value depends only on the characteristics of each individual.

We can thus write

$$w_{i,t} = \overline{w_t} + \Delta_{i,t}$$

where both $\overline{w_t}$ and $\Delta_{i,t}$ are stochastic variables. In particular, $\overline{w_t}$ is the average of $w_{i,t}$ taken with respect to females that make up the population at time t while, by definition of Δ, we have that $E[\Delta_{i,t}] = 0$, i.e. the expected value of the deviations vanishes. It is customary to assume that the distribution of individual deviations does not depend on time, that is $\Delta_{i,t} = \Delta_i$. Basically, the individual fitness in a breeding season is thus the sum of a random variable that reflects the effect of environmental stochasticity ($\overline{w_t}$) and one that accounts for demographic stochasticity (Δ_i). Figure 3.7 provides an idea of the variation of individual fitnesses—and particularly the diversification of the value of $\overline{w_t}$ and the values of Δ_i—in two populations of birds: the song sparrow (*Melospiza melodia*) and the great tit (*Parus major*). The contributions of the two different sources of stochasticity to the finite growth rate can be calculated as

$$\lambda_t = \frac{1}{N_t} \sum_{i=1}^{N_t} w_{i,t} = \overline{w_t} + \frac{\sum_{i=1}^{N_t} \Delta_i}{N_t}. \tag{3.1}$$

In conclusion, the random variable λ_t—remember that it is the average fitness in year t—is then the sum of two random variables, one depending on the year (environmental stochasticity) and the other depending on the variability between individuals (demographic stochasticity). Therefore, it is reasonable to assume that the two random variables are independent.

The simplest statistical properties of λ_t can be obtained by assuming that environmental stochasticity is a stochastic process without a *trend*, i.e. it is stationary. In other words, we assume that the mean and variance of the process $\overline{w_t}$ are independent

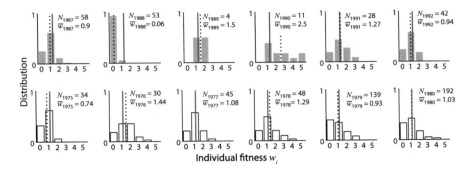

Fig. 3.7 Annual change in the distribution of individual fitnesses in two species of passerine birds: grey bars report the figures for the song sparrow (*Melospiza melodia*) at Mandarte Island, Canada, and white ones those of the great tit (*Parus major*) in Wytham Wood, England (after Lande et al. 2003). The dashed lines display the average fitness of the population in each year t (in the main text indicated as $\overline{w_t}$), while the solid line is the mean value of the average fitnesses across years (in the main text indicated as $\overline{\overline{w}}$). The number of sampled females from which statistics are derived is denoted by N_t

of time, namely

$$\mathrm{E}\left[\overline{w_t}\right] = \text{constant} = \overline{\overline{w}}$$
$$\mathrm{Var}\left[\overline{w_t}\right] = \text{constant} = \sigma_e^2.$$

If we denote by σ_d^2 the demographic variance—i.e. $\mathrm{Var}\left[\Delta_i\right]$—and keep in mind that $\overline{w_t}$ and Δ_i can be supposed to be independent random variables (because $\overline{w_t}$ depends on the variability of environmental conditions, while Δ_i depends on the variability of individuals), we obtain from Eq. 3.1

$$\mathrm{E}\left[\lambda_t\right] = \overline{\overline{w}}$$
$$\mathrm{Var}\left[\lambda_t\right] = \sigma_\lambda^2 = \mathrm{Var}\left[\overline{w_t}\right] + \frac{1}{N_t^2}\sum_{i=1}^{N_t}\mathrm{Var}\left[\Delta_i\right] = \sigma_e^2 + \frac{\sigma_d^2}{N_t}.$$

Table 3.2 shows that the demographic variance is in the order of 0.1–1 for many populations, while the environmental variance is usually one order of magnitude smaller. However, demographic stochasticity influences the variance of the growth rate through the factor $\frac{1}{N_t}$ and therefore it is practically immaterial for large populations, namely whenever

$$N \gg \frac{\sigma_d^2}{\sigma_e^2}.$$

Table 3.2 Demographic variance ($\hat{\sigma}_d^2$) and environmental variance ($\hat{\sigma}_e^2$) in populations with different ages at first reproduction (α). Bibliographic references and data after Table 1.2 in Lande et al. (2003)

Species	Location	α	$\hat{\sigma}_d^2$	$\hat{\sigma}_e^2$
Barn swallow (*Hirundo rustica*)	Denmark	1	0.18	0.024
White-throated dipper (*Cinclus cinclus*)	Southern Norway	1	0.27	0.21
Great tit (*Parus major*)	Wytham Wood, U.K.	1	0.57	0.079
Pied flycatcher (*Ficedula hypoleuca*)	Hoge Veluwe, The Netherlands	1	0.33	0.036
Song sparrow (*Melospiza melodia*)	Mandarte Island, Canada	1	0.66	0.41
Soay sheep (*Ovis aries*)	Hirta Island, U.K.	1	0.28	0.045
Brown bear (*Ursus arctos*)	Southern Sweden	4	0.16	0.003
Brown bear (*Ursus arctos*)	Northern Sweden	5	0.18	0.000

In an empirical way, for a given population we can define a critical population number

$$N_c = 10 \frac{\sigma_d^2}{\sigma_e^2}$$

below which demographic stochasticity cannot be neglected. The critical number N_c is often in the order of hundreds, but can indicatively vary between 10 and 1000, as one can check by calculating N_c for the populations listed in Table 3.2.

3.4 Risk Analysis for Populations Subject to Demographic Stochasticity Only

In this section we consider small populations (i.e. for which $N < N_c$) so that as a first approximation one can neglect environmental but not demographic stochasticity. Just because populations are very small, one can often assume that the rate of demographic growth does not depend on density and dynamics is therefore Malthusian. To facilitate understanding of the main concepts that guide the analysis, it is convenient to assume population dynamics in continuous rather than discrete time. However, the results would not change qualitatively if one considered the mathematically more complex case of discrete reproduction. Recall that the instantaneous birth-rate ν and death-rate μ can be reinterpreted in terms of individual fitness as

νdt = probability that a female produces one daughter in the small time interval dt

μdt = probability for a female to die in the small time interval dt.

The analytical approach to demographic stochasticity is not easy. We simply summarize the most important results in the Malthusian case (ν = constant, μ = constant, both independent of N) and then briefly treat the case with density dependence. The reader can refer to Iannelli and Pugliese (2014) for a more thorough analysis.

3.4.1   Demographic Stochasticity in Malthusian Populations

We denote by $N(t)$ the population abundance (more precisely, the number of females) at time t keeping in mind that, as the population is small, N must be considered as an integer, not a real number, because actually each organism is a discrete entity. It should also be noted that, although the population initially consists of exactly N_0 individuals, the abundance $N(t)$ at time $t > 0$ is a random variable. The simplest question we can ask about this random variable is how its average changes over time. The average is defined as

$$E[N(t)] = \overline{N}(t) = \sum_{i=0}^{\infty} i\, p_i(t) = \sum_{i=1}^{\infty} i\, p_i(t)$$

where $p_i(t)$ is the probability that at time t the population consists of an integer number i of individuals. The result that stems from the theory of stochastic processes (see e.g. Pielou 1977) is extremely simple, because the average population size satisfies the equation

$$\frac{d\overline{N}}{dt} = (\nu - \mu)\,\overline{N} = r\overline{N}$$

which is the same equation that governs the dynamics of the deterministic Malthusian model. Therefore, it turns out

$$\overline{N}(t) = \overline{N}(0)\exp(rt) = N_0 \exp(rt)$$

which implies that the average population size increases or decreases over time in accordance with the instantaneous rate of growth $r = \nu - \mu$ being positive or negative. Note that the average must be calculated from the infinite possible time evolutions (the realizations of the stochastic process) that can be followed by the population starting from the initial condition N_0.

However, one must not think that nothing changes with respect to the deterministic model. In particular, this result does not imply that the population can become extinct only if the growth rate r is negative. After some non-elementary calculations starting from an infinite system of nonlinear differential equations (the so-called Kolmogorov equations) in the variables $p_i(t)$—i.e. the probabilities that the population be composed of exactly $i = 0, 1, 2, \ldots$ individuals—one gets the most important result of the theory of demographic stochasticity. This result concerns the dynamics over time t of the probability $p_0(t)$ that the population be composed of zero individuals, or in other words of the *extinction risk*. One can in fact prove (Iannelli and Pugliese 2014) that the time dynamics of p_0 is given by the formula

$$p_0(t) = \left(\frac{\mu \exp(rt) - \mu}{\nu \exp(rt) - \mu} \right)^{N_0} = \left(\frac{\mu \exp(rt) - \mu}{(r + \mu) \exp(rt) - \mu} \right)^{N_0}, \qquad (3.2)$$

where N_0 is the initial number of adult females (for simplicity we can assume that the sex ratio remains constant over time and does not affect the birth rate).

First of all, we note that the extinction probability is positive even for $r > 0$. This is a fundamental result: demographic stochasticity can lead to extinction even Malthusian populations with a positive growth rate. Secondly, we can calculate the extinction probability of the population as t tends to infinity. This calculation allows us to understand how likely the population is to die out in the long run. Obviously, the asymptotic extinction risk depends on the value of parameter r. Suppose first that $r < 0$; one obtains

$$\lim_{t \to \infty} p_0(t) = \left(\frac{-\mu}{-\mu} \right)^{N_0} = 1.$$

So in the long run extinction is certain for populations with negative growth rate. The result is indeed a foregone conclusion.

Suppose now that $r > 0$. It is easy to get

$$\lim_{t \to \infty} p_0(t) = \lim_{t \to \infty} \left(\frac{\mu \exp(rt) - \mu}{\nu \exp(rt) - \mu} \right)^{N_0} = \left(\frac{\mu}{\nu} \right)^{N_0} = \left(\frac{\mu}{r + \mu} \right)^{N_0}.$$

Therefore, in the long run, there exists a non-vanishing extinction probability even for populations that on the average are growing, as shown in Fig. 3.8.

However, it is worth noting that a long time might be necessary before the extinction probability becomes close to the long-term value. Recalling that $\frac{1}{\mu}$ is the average lifetime of an organism belonging to the population—and is also the average generation length in the case of semelparous populations—we can wonder what the risk of extinction is after a reasonable time, for instance 10 or 50 generations. Calculations are easy thanks to Eq. 3.2. Table 3.3 reports the results of computations that show that, for high mortality and small (albeit positive) growth rate, the probability of extinction after 50 generations is not very different from the asymptotic one. It

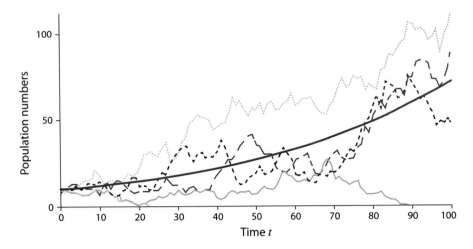

Fig. 3.8 Two (solid yellow and green) out of five stochastic simulations of the Malthusian process shown in figure terminate in extinction, although the growth rate is $r = 0.02$ time-unit^{-1}. In red bold, the trend of the theoretical average $\overline{N}(t)$

Table 3.3 Extinction probability for a Malthusian population characterized by $r = 0.05$ time-unit^{-1} and $\mu = 0.5$ time-unit^{-1} and initially composed by N_0 females: asymptotic probability (second column), probability after 10 and 50 generations (third and fourth column)

N_0	Asymptotic Prob.	10 Gener. Prob.	50 Gener. Prob.
1	0.909	0.863	0.908
5	0.621	0.480	0.619
10	0.386	0.230	0.383
50	0.00852	0.000647	0.00826
100	7.26×10^{-5}	4.18×10^{-7}	6.82×10^{-5}
500	2.01×10^{-21}	1.28×10^{-32}	1.48×10^{-21}

should be noted that the extinction probability is anyway very small for populations with more than 50 females (that is, 100 individuals if the sex ratio is 1:1).

Finally, we can calculate the extinction probability for stationary populations ($r = 0$, namely they are stationary from a deterministic viewpoint). This is a bit critical, because we must resort to *de L'Hôpital* rule for indeterminate forms:

$$\lim_{r \to 0} p_0(t) = \lim_{r \to 0} \left(\frac{\mu t \exp(rt)}{(\mu + r)t \exp(rt) + \exp(rt)} \right)^{N_0} = \left(\frac{\mu t}{1 + \mu t} \right)^{N_0}.$$

We can then easily calculate the asymptotic extinction probability

$$\lim_{t \to \infty} p_0(t) = \lim_{t \to \infty} \left(\frac{\mu t}{1 + \mu t} \right)^{N_0} = 1.$$

Table 3.4 Extinction probability at different times t in stationary populations ($r = 0$, death-rate = μ) for increasing initial population size N_0

N_0	$t = 1/\mu$	$t = 10/\mu$	$t = 100/\mu$	$t = 1000/\mu$
1	0.5	0.909	0.990	0.999
5	0.031	0.621	0.951	0.995
10	0.00097	0.386	0.905	0.990
50	8.88×10^{-6}	0.0085	0.608	0.951
100	7.89×10^{-31}	0.000072	0.370	0.905
500	~ 0	2.01×10^{-21}	0.0069	0.607

Therefore, long-term-extinction is certain in stationary populations, even if the average value of $N(t)$ is always constant and equal to the initial population N_0. This may sound like a paradox, but basically means this: if we replicated the demographic process very many times with initial population size equal to N_0, in almost all cases the population would become extinct sooner or later, but in some very rare cases the population would succeed in avoiding small numbers and grow exponentially thus keeping the average size constant. Note that, if N_0 is large, $p_0(t)$ can tend to 1 very slowly (see Table 3.4). Anyway, after 1000 generations the extinction probability of a stationary population starting from 500 females is about 60%.

It is also interesting to see how long it takes on average for a population to become extinct. A very simple index is the median time to extinction. As the time to extinction is a stochastic variable, then $p_0(t)$ is also the probability that the time to extinction be $\leq t$. Therefore the median time to extinction t_{med} is the one at which $p_0(t_{\text{med}}) = 0.5$. With easy calculations one gets for stationary populations ($r = 0$)

$$t_{\text{med}} = \frac{1}{\mu} \frac{1}{2^{\frac{1}{N_0}} - 1} = \frac{1}{\mu} \frac{1}{\exp\left(\frac{\ln 2}{N_0}\right) - 1} \simeq \frac{1}{\mu} \frac{1}{\frac{\ln 2}{N_0} + 1 - 1} = \frac{1}{\mu} \frac{N_0}{\ln 2}.$$

So with $N_0 = 5$, the median extinction time of a stationary population is about 7 generations, with $N_0 = 50$ it is approximately 70 generations and with $N_0 = 500$ about 700 generations. If the population is non-stationary, one must distinguish between decreasing ($r < 0$) and increasing ($r > 0$) populations. In decreasing populations, the asymptotic probability of extinction is one. Therefore, using the formula given by Eq. 3.2 and equating the left-hand-side to 0.5, one can solve with respect to time and obtain a general expression for the median time to extinction for decreasing populations (the derivation is left to the reader as an exercise):

$$t_{\text{med}} = \frac{1}{r} \ln \left(\frac{2^{\frac{1}{N_0}} - 1}{2^{\frac{1}{N_0}} - 1 - \frac{r}{\mu}} \right).$$

With $r = -0.05$ time-unit^{-1}, the time to extinction is 6 generations for $N_0 = 5$, 30 generations for $N_0 = 50$, and 42 generations for $N_0 = 100$.

In increasing population, only the fraction $\left(\frac{\mu}{\nu}\right)^{N_0}$ of long-term possible time evolutions will become extinct. The remaining fraction will increase indefinitely. So the median time to extinction must be calculated conditional on population extinction. The probability for the population to become extinct within time t conditional on extinction is $p_0(t)/(\mu/\nu)^{N_0}$. By equating this expression to 0.5, one can obtain, after boring calculations, the following formula

$$t_{\text{med}} = \frac{1}{r} \ln \left(\frac{2^{\frac{1}{N_0}} - 1 + \frac{r}{\mu} 2^{\frac{1}{N_0}}}{\left(2^{\frac{1}{N_0}} - 1\right)\left(1 + \frac{r}{\mu}\right)} \right).$$

With $r = 0.02$ time-unit^{-1}, the conditional median time to extinction is 6 generations fo $N_0 = 5$, 44 generations for $N_0 = 50$, and 67 generations for $N_0 = 100$.

3.4.2   Demographic Stochasticity in Density-Dependent Populations

As the landscape is currently very much fragmented, it may well occur that the habitat of and available resources for wildlife populations are greatly reduced. Therefore, intraspecific competition can operate even in small populations for which it is mandatory to include demographic stochasticity in order to conduct a correct risk evaluation. The theory just illustrated can be adapted to the case where the rate ν and μ are not constant (as in the Malthusian growth model) but are dependent on the number N of organisms. For example, one can consider the case in which the dependence on density is of the logistic type, namely

$$\nu_N = \nu_0 - aN$$
$$\mu_N = \mu_0 + bN$$

with ν_0, μ_0, a and b being positive constants. The per capita instantaneous growth rate is thus given by

$$r_N = \nu_0 - \mu_0 - (a + b)N = r_0 - cN.$$

It is possible to prove (Iannelli and Pugliese 2014) that, if there is density dependence, extinction is always certain, which means that in any case it turns out that

$$\lim_{t \to \infty} p_0(t) = 1.$$

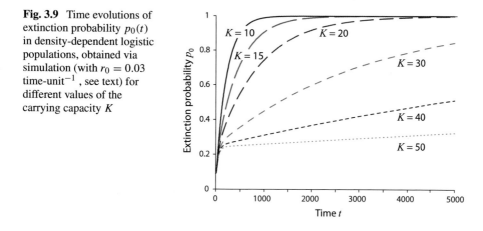

Fig. 3.9 Time evolutions of extinction probability $p_0(t)$ in density-dependent logistic populations, obtained via simulation (with $r_0 = 0.03$ time-unit^{-1} , see text) for different values of the carrying capacity K

The case is in a sense analogous to the Malthusian one with stationary populations. In fact, in the deterministic logistic model the growth rate vanishes in correspondence to the carrying capacity, which is the equilibrium toward which the deterministic logistic growth tends. In the stochastic logistic model, we can still introduce a carrying capacity $K = r_0/c$, but the time evolutions of the population, after possibly oscillating randomly around K, will go to zero sooner or later with probability one. However, if the carrying capacity is large, $p_0(t)$ tends to 1 very slowly, as shown in Fig. 3.9. Indeed, this figure clearly demonstrates the so-called phenomenon of the elbow curve. For carrying capacities larger than about 40 individuals, the time development of $p_0(t)$ is divided into two phases: the first phase is a rapid increase over time of the extinction probability, while in the second phase the extinction probability increases so slowly that even after hundreds of generations it turns out to be much smaller than the theoretical asymptotic value, which is equal to 1. The carrying capacity of each habitat fragment thus plays a very important role in determining the actual risk of extinction.

3.5 Risk Analysis for Populations Subject to Environmental Stochasticity Only

For populations whose size is not too small (say $N_c \ll N$) we can neglect demographic stochasticity and consider the environmental one only. Suppose then that individual fitnesses are all equal, but time-dependent

$$w_{i,t} = \overline{w_t}.$$

Since the population numbers are not too small, we can approximate N as a real, not an integer variable and thus write

$$N_{t+1} = \sum_{i=1}^{N_t} w_{i,t} = \overline{w_t} N_t = \lambda_t N_t.$$

The finite growth rate λ_t can differ from year to year due to changes in the external environment (climate, physical and chemical conditions in soil or water, the presence of predators, etc.). These changes affect the average fertility and/or survival of the organisms and thus indirectly affect the rate of population growth. The model we will use to describe environmental stochasticity will always be of this type

$$\lambda_t = \Lambda(N_t)\delta_t$$

namely

$$N_{t+1} = \delta_t \Lambda(N_t) N_t \qquad (3.3)$$

where δ_t is a random factor. In other words, we assume that environmental variability influences the rate of demographic growth in a multiplicative way. δ_t is called environmental multiplicative noise and is a number ≥ 0, so as to guarantee that the model always provides non-negative abundances N_t. As we assume that all years are mutually independent and that there is no environmental trend, δ_t is a random variable that is always drawn from the same probability distribution. For example, many data suggest that δ_t is often distributed as a lognormal—i.e. $\log(\delta_t)$ is distributed as a normal variable. It should be noted that the deterministic model

$$N_{t+1} = \Lambda(N_t) N_t$$

can be obtained from model (3.3) by setting the multiplication factor δ_t to 1. It is convenient to assume that $\delta_t = \exp(\varepsilon_t)$ with ε_t "*white noise*". A stochastic process is called white when it has the following properties

$$E[\varepsilon_t] = 0$$
$$E[\varepsilon_t^2] = \mathrm{Var}[\varepsilon_t] = \sigma_\varepsilon^2$$
$$E[\varepsilon_t \varepsilon_{t-\tau}] = 0 \text{ for } \tau \neq 0.$$

The last property implies no correlation between the different years. With these assumptions, the median of the random variable δ_t (the value corresponding to the 50-th percentile) is equal to 1. The deterministic model can therefore be thought of as the median model of the situation with environmental stochasticity. As stated before, it is reasonable to assume that ε_t be normal and thus δ_t lognormal. In this case, the graph of the probability density of the multiplicative noise is as shown in Fig. 3.10. It should be strongly remarked that while the random variable ε, which is normally distributed, has the property that mean, mode and median coincide, this is not the case for the variable $\delta = \exp(\varepsilon)$. Only the median is preserved after exponential transformation, or

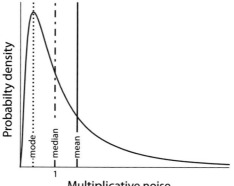

Fig. 3.10 Lognormal probability density. The relevant parameters are, respectively, the mean μ_ϵ and standard deviation σ_ϵ of the underlying normal random variable. In this graph $\mu_\epsilon = 0$ and $\sigma_\epsilon = 1$. The dotted, dash-dotted and solid segments, respectively, demarcate the mode, the median and the mean of the distribution. Since $\mu_\epsilon = 0$, the median of the lognormal is equal to 1

$$\text{Median of } \delta = \exp(\text{Median of } \varepsilon) = \exp(0) = 1.$$

The mode of δ, instead, is smaller than 1 and the mean larger than 1. In particular, one can prove that

$$E[\delta] = \exp\left(\frac{\sigma_\varepsilon^2}{2}\right)$$

where σ_ε^2 is the variance of ε. Notice that the variance of the environmental logarithmic noise (σ_ε^2) should not be confused with the variance of $\overline{w_t}$ that we previously indicated by σ_e^2.

3.5.1 Environmental Stochasticity in Malthusian Population

In Malthusian populations $\Lambda(N_t) = \lambda$ constant so that

$$N_{t+1} = \lambda N_t \exp(\varepsilon_t).$$

Note that λ is actually the median rate of growth (in fact it can be obtained by setting $\varepsilon = 0$, i.e. $\delta = 1$). The corresponding deterministic model is

$$N_{t+1} = \lambda N_t$$

and then—if there were no stochasticity—the population would tend to become extinct if $\lambda < 1$. If environmental stochasticity is introduced, the growth rate varies from one year to the next. Therefore, given an initial condition with abundance equal to N_0, the time evolution of the population size is

$$N_t = \lambda_{t-1}\lambda_{t-2}...\lambda_0 \cdot N_0 = \lambda^t \exp(\varepsilon_0 + \varepsilon_1 + ... + \varepsilon_{t-1}) \cdot N_0.$$

To understand the long-run trend of abundance we consider the dynamics of N_t in logarithmic scale, namely

$$\log N_t = t \log \lambda + \log N_0 + \psi_t = rt + \log N_0 + \psi_t \tag{3.4}$$

where $\psi_t = \sum_{i=0}^{t-1} \varepsilon_i$ is an additive noise given by the sum of the ε_t while $r = \log \lambda$. It is easy to deduce the properties of ψ_t, because

$$E[\psi_t] = \sum_{i=0}^{t-1} E[\varepsilon_i] = 0 \tag{3.5}$$

$$\text{Var}[\psi_t] = \sum_{i=0}^{t-1} \text{Var}[\varepsilon_i] = t\,\text{Var}[\varepsilon_i] = t\sigma_\varepsilon^2. \tag{3.6}$$

The relationship (3.6) stems from the random variables ε_t being statistically independent of each other (white noise property).

It is worthwhile to remark that, while the mean value of ψ_t is null, its variance increases linearly with time. In conclusion, by considering abundances in logarithmic scale ($Z_t = \log(N_t)$) one gets

$$Z_t = t \log \lambda + Z_0 + \psi_t = rt + Z_0 + \psi_t$$
$$E[Z_t] = t \log \lambda + Z_0 = rt + Z_0$$
$$\text{Var}[Z_t] = \text{Var}[\psi_t] = t\sigma_\varepsilon^2.$$

In particular, if ε_t is normally distributed, ψ_t is normal too, because it is the sum of normal and independent variables. Therefore, the logarithmic abundance is distributed as a Gaussian and its average increases (or decreases) linearly with time, while its standard deviation increases with the root of time (see Fig. 3.11). We thus conclude that a population subject to environmental stochasticity tends to grow on average if $r = \log \lambda > 0$ (i.e., if $\lambda > 1$) just like in the deterministic model. However the variance of the logarithm of abundance also tends to grow over time. So, even with $\lambda > 1$, for increasing times, there can occur very small population sizes with high probability (again see Fig. 3.11).

It must be strongly remarked that λ is the *median*, not the mean, value of λ_t. To establish whether a population is growing or declining on the average, one might be tempted to calculate $\lambda_t = \frac{N_{t+1}}{N_t}$ from data and then find the mean of the λ_t and see if it is larger or smaller than 1. This procedure is however incorrect. In fact, we have already seen that the lognormal distribution of multiplicative noise δ_t (or equivalently the normal distribution of ε_t) implies

Fig. 3.11 Distributions, at various times ($t = 0, 20, 100, 200$ and 400 years), of the natural logarithm of a hypothetical population size obtained by simulation from the initial condition $\log(N_0) = 5$. The assumption is Malthusian growth in the presence of environmental stochasticity only (no demographic stochasticity). Parameter values are $r = 0.05\,yr^{-1}$ and $\sigma_\epsilon^2 = 0.05$. Redrawn after Lande et al. (2003)

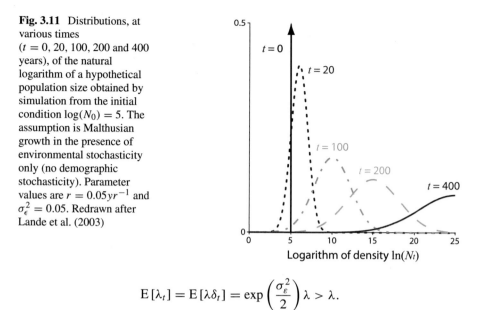

$$E[\lambda_t] = E[\lambda \delta_t] = \exp\left(\frac{\sigma_\varepsilon^2}{2}\right)\lambda > \lambda.$$

Therefore a population can be declining on average ($\lambda < 1$) even if the mean value of the finite rates of increase (as calculated from yearly counts) is larger than 1. The likelihood of this event is larger if environmental stochasticity is higher (larger σ_ε^2) because the multiplication factor $\exp(\frac{\sigma_\varepsilon^2}{2})$ can be much greater than unity. If we applied the incorrect estimate, we might falsely envisage a rosy future for a population by estimating its growth rate as being greater than 1, whereas in reality a correct estimate would provide a value less than 1. Note that we can instead correctly use the average of the logarithmic growth-rates. Indeed

$$E\left[\log \lambda_t\right] = E\left[\log(\lambda \delta_t)\right] = \log \lambda + E\left[\log \delta_t\right] = r + E[\varepsilon_t] = r.$$

To illustrate this concept in an effective manner by means of an example, Lewontin and Cohen (1969) present the hypothetical case of a population with annual reproduction whose growth-rate is influenced by environmental conditions. In particular, they assume that usually the growth rate is $\lambda = 1.1$, however there are some critical years in which $\lambda = 0.3$. These critical years occur with low probability, let's say on average once every ten years. Thus the average of the λ_t would amount to $1.1 \times \frac{9}{10} + 0.3 \times \frac{1}{10} = 1.02$. One might falsely conclude that the expected value of the abundance of the population would be growing by an average of 2% annually. Instead, by calculating the average of the logarithmic rates of increase one gets $\log 1.1 \times \frac{9}{10} + \log 0.3 \times \frac{1}{10} = -0.0346$ which implies that the instantaneous growth rate r is negative. The correct estimate of λ is therefore $\exp(-0.0346) = 0.966 < 1$, which leads to the conclusion that the population is doomed to extinction in the long term. Figure 3.12 shows, on a logarithmic scale, the abundances obtained via 100 simulations for a population whose growth rate is randomly drawn each year and is just 1.1 with probability 90% and 0.3 in

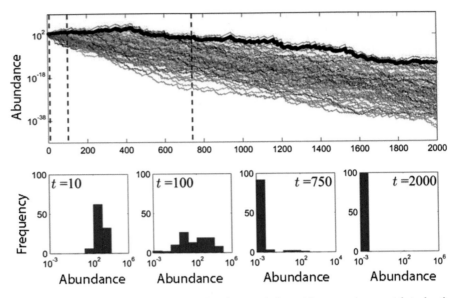

Fig. 3.12 Possible time evolutions of a Malthusian population subject to environmental stochasticity only. The top panel shows in logarithmic scale the time trend of population abundances for 100 repeated simulations starting from the same initial conditions ($N_0 = 100$), and their mean value (thick line). The bottom boxes show the histograms of frequencies of these abundances at four different times

the remaining 10% of the cases. As one can see from the time evolution of the average abundance, the population's fate is extinction. The bottom panels of the same figure display the histograms of abundance frequencies in subsequent generations (respectively $t = 10, 100, 750, 2000$). From these histograms it is apparent that in almost all simulations abundance is practically vanishing after 2,000 generations.

It is thus of utmost importance to correctly estimate λ from available data (e.g. from periodical population census, as those reported in some of the exercises at the end of this chapter). The proper way to address the problem is to not directly estimate λ, rather $r = \log \lambda$. To this purpose it is sufficient to use the standard linear regression. As suggested by Eq. 3.4, one must simply make a regression of $\log (N_t)$ against time t. The slope of the regression line provides a correct estimate $\hat{r}$ of r. If $\hat{r}$ is positive we can infer that the population tends to grow on average. Of course, the greater the number of data on which we base the estimate of r, the greater the confidence in this statement.

3.5.2   Estimating the Risk of Extinction

In the model we are considering, unlike the one we used to study demographic stochasticity, extinction is never possible in a theoretical sense because the variable N_t can never vanish in finite time if the initial condition N_0 is positive. However, we know that below the critical threshold N_c environmental stochasticity loses its importance and at the same time demographic stochasticity is no longer negligible and begins operating. In the same way, in cases where depensation or genetic deterioration plays a role, there exist abundance thresholds below which processes at the scale of each individual can strongly influence a population fate. It then becomes relevant to find the probability that N_t falls below a critical size N_c thus dragging the population into a so-called "vortex" of demographic stochasticity. To calculate this probability it is just sufficient to consider the logarithmic threshold $Z_c = \log(N_c)$ and calculate the size of the tail to the left of Z_c for the probability distribution of $Z_t = \log(N_t)$. We know that, if ε_t is normally distributed, Z_t is normal too and has average equal to $rt + Z_0$ and variance equal to $t\sigma_\varepsilon^2$. If we then knew r and σ_ε^2 we could easily estimate the probability for the population to fall below Z_c at time t (see Fig. 3.13).

Instead of the theoretical values r and σ_ε^2, which are generally unknown, we can use their estimates. We already learnt how to calculate $\hat{r}$, but we have to understand how to estimate the environmental variance σ_ε^2. This is less elementary. If we have m available data (all in consecutive years) one can proceed as follows (otherwise things are a bit more complicated). We know that

$$N_{t+1} = \lambda N_t \exp(\varepsilon_t)$$

and hence

$$\log\left(\frac{N_{t+1}}{N_t}\right) = \log\lambda + \varepsilon_t = r + \varepsilon_t.$$

If in the previous relationship we replace r with $\hat{r}$, we can get an estimate of the environmental noise as

$$\hat{\varepsilon}_t = \log\left(\frac{N_{t+1}}{N_t}\right) - \hat{r}$$

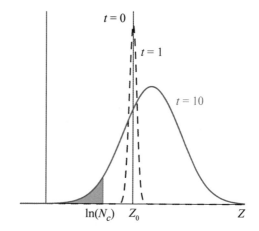

Fig. 3.13 Hypothetical distributions of the logarithm of the abundances for three different instants of time $t = 0, 1$ and 10 in a population subject to environmental stochasticity only. The area of the green shaded region corresponds to the probability for the population to drop below the critical threshold at time $t = 10$

and consequently estimate the environmental variance σ_ε^2 as

$$\widehat{\sigma}_\varepsilon^2 = \frac{1}{m-2} \sum_{t=1}^{m-1} \widehat{\varepsilon}_t^2$$

where the formula for non-biased variance estimation is used.

3.5.3 Environmental Stochasticity in Density-Dependent Populations

So far we have studied the effects of environmental stochasticity in Malthusian populations. This is not unreasonable because, if our main aim is to estimate the extinction risk in not too big populations, we can, as a first approximation, neglect the phenomena of intraspecific competition. However, we have previously seen that intraspecific competition is not the only mechanism of density dependence. Sometimes there may be density dependence effects even in relatively small populations (such as for example in the case of depensation). Also, if the habitat suitable for a species has been greatly reduced, there can be competition even in populations with numbers close enough to the threshold of demographic stochasticity or genetic deterioration. Therefore, in some cases it may be reasonable to consider a model like the following

$$N_{t+1} = \exp(\varepsilon_t)\Lambda(N_t)N_t$$

where $\Lambda(N_t)$ is the median growth rate, ε_t is the white noise that incorporates environmental stochasticity and $\Lambda(N_t)$ is not a constant, but depends on density. First it is to be noted that

$$\log\left(\frac{N_{t+1}}{N_t}\right) = \log \lambda_t = \log \Lambda(N_t) + \varepsilon_t$$

and thus, in order to estimate $\Lambda(N)$, one can make a regression (possibly non-linear) of $\log \lambda_t$ against N_t. The resulting regression curve can be taken as an estimate of $\log \Lambda(N)$ from which one can derive $\Lambda(N)$.

The properties of the deterministic model depend on the functional form of $\Lambda(N)$. If only intraspecific competition is operating, $\Lambda(N)$ is a decreasing function of N with $\Lambda(0) > 1$. Typical examples are the Beverton-Holt and Ricker models already mentioned above. For example, Fig. 3.14a shows the time evolution of the red deer population in the Yellowstone Park, USA. Clearly growth is not Malthusian, but density dependent. We can test whether the Ricker model provides a good description of available data. To this end we note that the Ricker model can also be written as

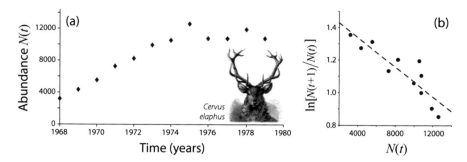

Fig. 3.14 **a** Time evolution of the population of red deer (*Cervus elaphus*) in the Yellowstone Park, and **b** logarithmic growth rates versus deer abundance (filled circles) and fitted linear regression line (dash). Reworked after Lande et al. (2003)

$$\log\left(\frac{N_{t+1}}{N_t}\right) = \log\lambda - \beta N_t + \varepsilon_t$$

and thus—apart from the environmental noise ε_t—the logarithmic growth-rates are related to density N_t via a linearly decreasing relationship. In fact Fig. 3.14b shows that as a first approximation a straight line interpolates the data quite well. Using the classic linear regression formulas we obtain the following estimates of the demographic parameters: $\lambda = 1.597$, $\beta = 4.135 \times 10^{-5}$.

At present it is difficult, if not impossible, to theoretically analyse the properties of stochastic models with density dependence. In almost all cases we must resort to extensive computer simulation. Usually a fixed time horizon (for example, 100 years) is chosen and a number of simulations (for example 500) is carried out from given initial conditions. One can then calculate various indices, such as for example:

- the mean, standard deviation and median of the population abundance as a function of time;
- the extinction probability as a function of time, estimated as the percentage of replicates that have fallen below the critical threshold N_c within time t;
- the mean and median extinction times, estimated as the mean and median of the times at which the population size reaches the critical threshold (of course only for replicates that become extinct).

As an example, consider a population whose fluctuations are well described by a stochastic Ricker model, that is

$$N_{t+1} = \lambda N_t \exp\left(-\beta N_t\right)\exp(\varepsilon_t) \tag{3.7}$$

with $\lambda > 1$. Figure 3.15a shows three simulations with a stochastic Ricker model in which $\lambda = 4$ and $\beta = 0.001$. The equilibrium of the deterministic model is equal to $\frac{\log\lambda}{\beta} \approx 1,386$ individuals. Even though the deterministic model is characterized by a stable equilibrium (with damped oscillations because the slope of the curve at equilibrium is $1 - \log\lambda \approx -0.386$), the environmental variance is so high as

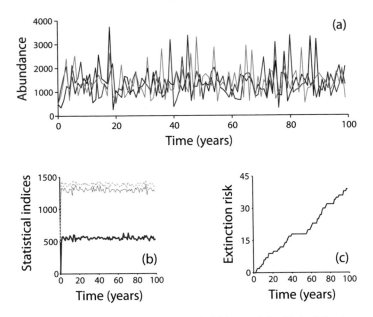

Fig. 3.15 Simulations obtained through the stochastic Ricker model, with the following parameters: $\lambda = 4$, $\beta = 0.001$, $\sigma_\epsilon = 0.35$. The deterministic equilibrium is $1,386$ individuals. **a** Time course of three simulations starting from 500 initial individuals. **b** Time evolution of the average (dotted yellow line), the median (thin black line) and standard deviation (thick pink line) of 500 simulations. **c** Time evolution of the extinction probability as a function of time, calculated as the fraction of simulations that fall below the threshold of 200 individuals

to produce permanent large fluctuations around this value. However, carrying out 500 simulations and reporting the time evolution of the mean, median and standard deviation of N_t, we see that the process is basically stationary, because each of these indicators tends to a constant value (Fig. 3.15b). Among other things, the mean tends to the deterministic equilibrium value. However, one should not be tempted to generalize this outcome, because this is a peculiarity of the Ricker model that does not apply to other models with density dependence (for istance, it does not apply to the Beverton-Holt model). In addition, one can note that the median abundance is always below the mean.

To calculate the extinction risk we must first set a threshold value as critical population size. Suppose, for example, that $N_c = 200$. The trend of the extinction probability is reported in Fig. 3.15c which shows that the risk of extinction within 100 years is approximately equal to 40%. For that 40% of simulations that become extinct, the average time to extinction is 53.7 years and the median time is 57 years. Of course, things would radically change if we considered a different critical threshold. If the threshold were to be reduced to 50 individuals the probability of extinction within 100 years would decrease down to 0.6%. Since one does not actually know very well the critical thresholds for the different species, it is always recommendable that extinction risk studies include a sensitivity analysis for varying extinction thresholds N_c.

3.6 Extinction Vortices, Population Viability Analyses, IUCN Risk Categories

In reality, the species that inhabit the earth are simultaneously subjected to the whole multiplicity of the extinction causes considered thus far: habitat destruction and fragmentation, overexploitation, genetic deterioration, global climate change, demographic and environmental stochasticity. Very often an initial reduction of a population size due for example to a catastrophic event (fire, spillage of pollutants, etc.), increases the risk of extinction due to another cause (e.g. competition with alien species) leading to a further reduction of the population abundance and thus making it more prone to being impacted by yet another cause of extinction (e.g. demographic stochasticity). This chain process has been called *extinction vortex* (Gilpin and Soulé 1986) in a somewhat fanciful but effective way and is graphically represented in Fig. 3.16.

To conduct a sufficiently realistic risk analysis for a population it is necessary to include all the factors that, to our knowledge, may affect its future dynamics. The techniques used for this purpose go generally under the name of Population Viability Analysis (PVA, Boyce 1992) and are often coded into specially developed computer programs, such as VORTEX, RAMAS, ALEX, NEMESIS. Currently the term PVA is used to describe the set of statistical tools and simulation techniques that allow the estimation of the probability that a population or group of populations may persist for a certain time in a given territory. The example in Fig. 3.17 reports the results of

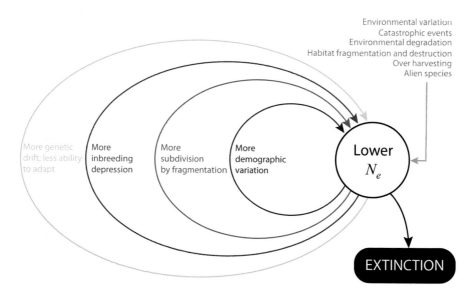

Fig. 3.16 Graphical representation of the so-called vortex of extinction. When a species enters the vortex, a sequence of phenomena gradually reduces a population size down to possible local extinction. Reworked after Primack (2000)

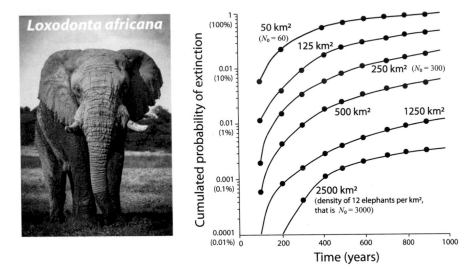

Fig. 3.17 Cumulated probability of extinction as a function of time for populations of African elephant (*Loxodonta africana*) in reserves of various sizes. The scale is semi-logarithmic. The initial elephant density is supposed to be anyway equal to 12 elephants per $10\,km^2$. For instance, an elephant population in a $125\,km^2$ protected area has a 40% probability of becoming extinct in 700 years. Modified after Armbruster and Lande (1993)

Table 3.5 Mace and Lande's (1991) criteria for the classification of species endangered by extinction. A species is placed into one of three categories (Critically Endangered, Endangered, Vulnerable) if it meets any one of the five criteria. N indicates the total population size, while N_s specifies the maximum size of subpopulations that make up the population, if it is fragmented

Demographic characteristic	Critical	Endangered	Vulnerable
Observed decline	80% in 10 years or 3 generations	50% in 10 years or 3 generations	20% in 10 years or 3 generations
Geographical range	$<100\,km^2$, single location	$<5000\ km^2$, <5 locations	$<20000\,km^2$, <10 locations
Total population	$N < 250$ $N_s < 50$	$N < 2,500$ $N_s < 250$	$N < 10,000$ $N_s < 1000$
Projected decline	>25% in 3 years or 1 generation	>20% in 5 years or 2 generations	>20% in 10 years or 3 generations
Extinction probability	>50% in 10 years or 3 generations	>20% in 20 years or 5 generations	>10% in 100 years

a PVA conducted for a critically endangered species: the African elephant. The goal of a PVA is often linked to management; in this case the authors of the analysis have tried to provide an idea about the influence of the size of possible reserves so that they can better protect the elephant populations.

The accuracy and reliability of a PVA critically depends on the available data regarding demographic and genetic parameters and the spatial distribution of individuals, the availability of present and future habitats, the influence of climatic factors, and so on. In most cases, the available data are extremely scarce, also because the species at greatest risk of extinction are often rare and poorly studied. For this reason it has been necessary to propose, in addition to formal analyses such as PVA, more empirical and speditive methods which allow the classification of the risk status of a species. Therefore, Mace and Lande (1991) proposed classification methods that are based on the type of information that is realistically available. The criteria make use of different indices that can be measured or quantified in just a few years. Table 3.5 shows the proposal by Mace and Lande (1991) which essentially is based on indices of observed or predicted decline, on the observed or predicted distribution range, on the size of single or fragmented population, on the extinction probability within a given time interval as obtained from a formal PVA. These criteria form the basis for the classification of IUCN (International Union for Conservation of Nature, http://www.iucn.org/) which is published under the name of Red List. After extensive discussion the classification methods of Mace and Lande were slightly changed, making them substantially more flexible. The current classification scheme (see Fig. 3.18) and the pertaining methods can be found in the latest edition of the IUCN manual (IUCN 2012b).

Fig. 3.18 The different categories of the current species classification according to the IUCN criteria that are the basis of the so-called Red List. Redrawn after IUCN (2012a)

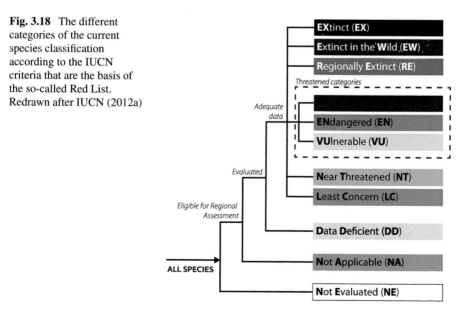

References

Armbruster P, Lande R (1993) A population viability analysis for African elephant (*Loxodonta africana*): How big should reserves be? *Conservation Biology* 7:602–610

Berger J (1992) Persistence of different-sized populations: An empirical assessment of rapid extinctions in bighorn sheep. *Conservation Biology* 4:91–98

Boyce N (1992) Population viability analysis. *Annual Review of Ecology and Systematics* 3:481–506

Burrows R, Hofer H, East ML (1995) Population dynamics, intervention and survival in African wild dogs (*Lycaon pictus*). *Proceedings of the Royal Society B: Biological Sciences* 262:235–45

Dennis B, Munholland P, Scott J (1991) Estimation of growth and extinction parameters for endangered species. *Ecological Monographs* 61:115–143

Gilpin ME, Soulé ME (1986) Minimum viable populations: processes of species extinction. In: Soulé ME (ed) *Conservation Biology: The Science of Scarcity and Diversity*. Sinauer Associates, Sunderland, MA, pp 19–34

Iannelli M, Pugliese A (2014) *An Introduction to Mathematical Population Dynamics*. Springer, Switzerland

IUCN (2012a) Guidelines for application of IUCN Red List criteria at regional and national levels: Version 4.0. Technical Report, International Union for Conservation of Nature, Gland, Switzerland and Cambridge, UK

IUCN (2012b) *IUCN Red List Categories and Criteria: Version 3.1*. IUCN, Gland, Switzerland and Cambridge, UK

Lande R, Engen S, Sæther BE (2003) *Stochastic Population Dynamics in Ecology and Conservation*. Oxford University Press

Lewontin RC, Cohen D (1969) On population growth in a randomly varying environment. *Proceedings of the National Academy of Science of the United States of America* 62:1056–1060

Mace G, Lande R (1991) assessing extinction threats: towards a reevaluation of IUCN threatened species categories. *Conservation Biology* 5:148–157

Parr R (1992) Nest predation and numbers of golden plovers *Pluvialis apricaria* and other moorland waders. *Bird Study* 40:37–41

Pielou E (1977) *Mathematical Ecology*. Wiley, New York, NY

Primack R (2000) *A Primer of Conservation Biology*. Sinauer Associates, Sunderland, MA, USA

Simmons R, Trewby I, Trewby M (1996) Are Etosha's blue cranes declining? *African Wildlife* 50:32–34

Chapter 4
Problems on the Analysis of Extinction Risk

Problem ER1

The Grünwald Park hosts a relict population of woolly bears (*Ursus laniger*). A team of German ecologists has been studying the demography of this rare (fantasy) species and finally has proposed the following discrete-time model:

$$N_{t+1} = L\,(N_t)\,N_t$$

where N is the bear population size and $L(N)$ is the finite rate of increase which depends on N via the following relationship

$$L(N) = \frac{6N}{480 + 0.8\,N + 0.01N^2}.$$

You are required to:

(a) qualitatively draw the cobweb plot for the population of *U. laniger* Grünwald Park;
(b) calculate the bear population sizes at equilibrium and assess their stability;
(c) assess the long-term fate of the population, given that the last count evaluated the bear population size to be 80 individuals.

Problem ER2

In an insect pest population with sexual reproduction the percentage of recessive homozygotes (genotype *aa*) is 36%. Under the assumption of Hardy-Weinberg proportions, calculate:

(i) The frequency of recessive and dominant alleles in the population
(ii) The frequency of dominant homozygotes and of heterozygotes
(iii) The frequency of the two possible phenotypes under the assumption that *A* is completely dominant with respect to *a*.

© The Author(s), under exclusive license to Springer Nature Switzerland AG 2022
M. Gatto and R. Casagrandi, *Ecosystem Conservation and Management*,
https://doi.org/10.1007/978-3-031-09480-4_4

Problem ER3 The Laysan
finch

Telespiza cantans

Problem ER3

The Laysan finch (*Telespiza cantans*) is a Hawaiian species classified as vulnerable by the IUCN. A genetic analysis conducted on 44 individuals evidenced a microsatelite locus (https://en.wikipedia.org/wiki/Microsatellite) with three alleles called 91, 95, 97. Below are the absolute frequencies of the genotypes in the 44 individuals sample

Genotypes	91/91	91/95	91/97	95/95	95/97	97/97	total
Numbers	7	10	8	5	11	3	44

(a) Derive the frequencies of the three alleles (91, 95, 97);
(b) Calculate the theoretical frequencies at Hardy-Weinberg equilibrium for the 6 genotypes, and
(c) Compare with the data from 44 finches.

Problem ER4

The common eider (*Somateria mollissima*) is a large sea-duck that is distributed over the northern coasts of Europe, North America and eastern Siberia. Here below we report the genetic data (Milne and Robertson 1965) of a small Scottish population; they pertain to two alleles (F and S) at an egg-white protein locus.

Genotypes	FF	FS	SS	Total
Numbers	37	24	6	67

(i) Calculate the frequencies of genes F and S and the genotypic frequency H of heterozygotes;

Problem ER4 The common
eider

(ii) Calculate the theoretical frequency H_{EQ} of heterozygotes that would establish
     if the population were actually at Hardy-Weinberg equilibrium; verify that
     $H_{EQ} > H$;
(iii) assume that the smaller frequency of heterozygotes is due to a bottleneck: at the
     very beginning the duck population was large, thus the heterozygote frequency
     was H_{EQ}; then the population has been drastically reduced so that the effective
     population size was 30 ducks for a certain time T, during which genetic drift
     has been operating; calculate the time T (years, because reproduction occurs
     once a year) that is necessary for letting heterozygosity go down to H.

Problem ER5

Wright's formula describes the decrease of heterozygosity due to genetic drift from
one generation to the next. Ryman et al. (1981) have studied three moose (*Alces*

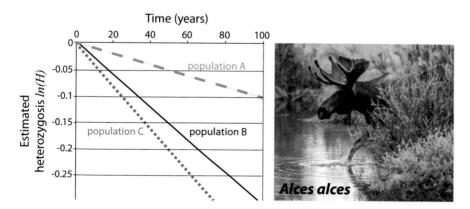

Problem ER5 Variation of moose heterozygosity in three Swedish populations, say A, B and C.
Modified after Ryman et al. (1981)

alces) populations in Sweden (henceforth indicated with the letters A, B, C) which have different generation times depending on the different hunting regulations that hold for each population. More precisely:

Population A: 9.9 years from one generation to the next
Population B: 7.5 years
Population C: 4.2 years.

The graph shows the time trend over a century of the estimated heterozygosity (natural logarithm of the proportion of heterozygous moose against time in years) for each of the three populations.

Calculate the effective population size for each of the three moose populations.

Problem ER6

The Iberian lynx (*Lynx pardinus*) provides a paradigmatic example of a species on the verge of extinction. There exist two remaining populations (Doñana and Andújar) that are isolated from one another. The population of Andújar is rather large while that of Doñana is small and subject to genetic deterioration. Casas-Marce et al. (2013) have studied the genetics of both populations. Heterozygosity was estimated to be 0.31 in Doñana and 0.46 in Andújar. Assume that the heterozygosity of the rather large population of Andújar is the original heterozygosity of Doñana, the one that was prevailing τ years ago before the Doñana population size was suddenly reduced to a low value.

Casas-Marce et al. (2013) have estimated that the effective population size N_e of Doñana is about 10 individuals. They also report that the generation length of the Iberian lynx is 5 years.

Estimate the value of τ, that is how many years ago the bottleneck occurred.

Problem ER6 The Iberian lynx

Problem ER7 The platypus

Ornithorhynchus anatinus

Problem ER7

The platypus, *Ornithorhynchus anatinus*, one of only five extant species of egg-laying mammals, is found in eastern Australia. Two island populations of *O. anatinus* exist there: King Island in Bass Strait, and Kangaroo Island off the coast of South Australia. The population on King Island is naturally occurring, while that on Kangaroo Island was established in 1941 by introducing a few individuals from a large population of the Victoria region. Kangaroo population is small and has therefore been subjected to genetic deterioration. Furlan et al. (2012) have studied the population: they have estimated that it consists of about 110 individuals and that its average heterozygosity in 2009 was 0.419. However, the effective population size is not known.

Calculate the effective population size on the basis of the following information:

(a) the 1941 heterozygosity can be assumed to be equal to 0.597, which is the current heterozygosity of the Victorian population;
(b) the platypus generation time can be assumed to be about 10 years.

Problem ER8

The European shag (*Phalacrocorax aristotelis*) is experiencing decline all over Europe. Velando et al. (2015) have studied the genetics of this bird in the Rias Baixas (Galicia, Spain) between 2004 and 2013.

They found a link between the finite rate of demographic increase and the homozygosity HL of shag populations (see figure). Homozygosity is simply the complement to one of heterozygosity. This relationship is well described as

$$\lambda(t) = 0.75 + 0.36 \exp\left(-(\mathrm{HL}(t)/0.5)^2\right),$$

where t is the year.

Problem ER8   The relationship between finite rate of demographic increase and homozygosity in a Spanish population of the European shag (on the right)

Suppose that in 2016 we have an isolated population of 50 shags that are subject to genetic drift only and whose demography is simply described from year to year by the rate λ. The heterozygosity in 2016 is 0.8. Only 40% of the shags reproduce each year. Using Wright's equation calculate the heterozygosity and the population size of birds in 2021.

Problem ER9

The dark red helleborine (*Epipactis atrorubens*) is an endangered perennial orchid. Hens et al. (2017) collected data from several small and large populations in Northern Finland.

(A) The estimated annual finite rate of demographic increase is reported in the figure. What is the demographic phenomenon that characterizes this species in Northern Finland?

(B) The same scientists have conducted a genetic analysis of the populations, using microsatellite markers. For the small population of Mataraniemi (sampled in 2013), whose effective population size can be considered constant in the past one hundred years or so, they have observed a heterozygosity H equal to 0.312. From gene frequencies they have calculated the theoretical Hardy-Weinberg heterozygosity H_0 which turns out to be 0.397. The duration of a generation can be estimated as the sum of the average length of the seed dormant stage (1.34 years) plus the average age of seedlings at reproductive maturity (7.4 years). Assume that 100 years before 2013 the population was still large and its heterozygosity was H_0, and then the effective population size was suddenly reduced to the present size N_e and stayed more or less constant for one century. Estimate the effective population size N_e.

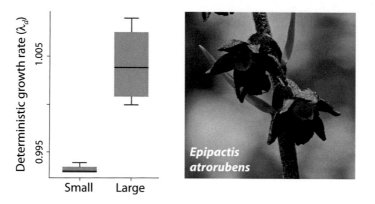

Problem ER9 The finite growth rate of the dark red helleborine (*Epipactis atrorubens*) for small ($N < 100$) and large ($N > 100$) populations. Redrawn after Hens et al. (2017)

Problem ER10

In 2023 a population of pink finches in the natural reserve of Greybeeches consists of only three adult reproductive females (from now on called A, B, C). They breed in mid spring. You know that:

- Female A produces 7 eggs, of which only 5 hatch; 3 nestlings are female and of them 2 will survive till adulthood and will reproduce in 2024; female A dies in December 2023;
- Female B produces 4 eggs, all of which hatch; 1 nestling is female but will not survive till adulthood; female B is still alive in mid spring 2024;
- Female C produces 9 eggs, of which only 6 hatch; 4 nestlings are female and of them 3 will survive till adulthood and will reproduce in 2024; female C is still alive in mid spring 2024;

Calculate the individual fitnesses of A, B e C and the finite rate of increase of the population between 2023 and 2024.

Problem ER11

It has been estimated that the brown bear population in southern Sweden is characterized by an environmental variance of 0.003 while the demographic variance is 0.16. Calculate the critical number of bears below which the population is captured by the vortex of demographic stochasticity.

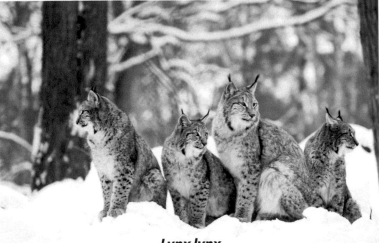

Lynx lynx

Problem ER12   The lynx

Problem ER12

The following table reports the numbers of a lynx (*Lynx lynx*) population in Sweden as censused during a decade

Year	1956	1957	1958	1959	1960	1961	1962	1963	1964	1965	1966
Numbers	11	8	3	10	10	18	11	21	19	55	45

Estimate the instantaneous rate of demographic increase for the Swedish lynx population. Suppose that you want to repopulate an Alpine region in which *Lynx lynx* is no longer present because it was exterminated in the past centuries. You decide to release 10 lynxes (sex ratio 1:1) which have been captured in Sweden. Calculate the long-term extinction risk for the new population. To this end assume that the average lifetime of a lynx is about 10 years and that the rate of demographic increase is the same as the one you estimated for the Swedish population.

Problem ER13

The bearded vulture (*Gypaetus barbatus*) is a bird of prey that eats mainly carrion and lives and breeds on crags in high mountains. Recently, it has been introduced again in the Alps (in particular in the Stelvio National Park, Italy). Schaub et al. (2009) report data on the numbers of vultures in the Alps between 1996 and 2006 (see table here below) and estimate that the birth rate (number of juveniles produced by a pair of vultures that will survive up to the reproductive age) is $0.6\,\text{year}^{-1}$.

Problem ER13 The bearded vulture

Gypaetus barbatus

Year	1996	1997	1998	1999	2000	2001
Number of pairs	1	1	2	4	4	5
Year	2002	2003	2004	2005	2006	
Number of pairs	6	6	8	9	9	

The reintroduction has been luckily successful because, given an initial population of two birds only in 1996, the vultures have now established in the Alps and the most recent estimate evaluates their population to have reached about one hundred individuals.

However, owing to demographic stochasticity, the reintroduction might have been less successful. Estimate how many vultures (males + females) should have been introduced into the Alps in order to have a long-term probability of success equal to 95%.

Problem ER14

Wilkinson and O'Regan (2003) report the following data for the populations of leopards (*Panthera pardus*) and tigers (*Panthera tigris*) in Indonesian islands:

- Females reproduce once a year with a variable litter size (number of male and female cubs) according to the table

Litter	1	2	3	4	5	6
Leopard (%)	30	60	10	–	–	–
Tiger (%)	10	38	38	10	3	1

- The sex ratio at birth is 1:1;
- 50% of both tiger and leopard cubs die before reaching adulthood;
- The average age at death for both tigers and leopards is about 10 years.

These big cats are now quite rare. Suppose you want to repopulate one of the Indonesian islands with leopards and tigers. How many pairs should you release in order to achieve a 95% probability of success in the long run?

Problem ER14 The big cats of Indonesia: leopard and tiger

Problem ER15

A population of spotted deer (*Dama variegatus*, a fantasy species) is driven by environmental stochasticity only. Its median finite rate of increase is $\lambda = 1.03$ while the environmental variance σ_ε^2 (namely the variance of the logarithm of the multiplicative noise that influences the population abundance) is 0.025. In 2022, 210 deer have been counted. How will the population abundance be distributed in 2032 if the growth is Malthusian?

Problem ER16

In Europe, the abandonment of agricultural activities and traditional forest usages have led to forest spread, which eventually affects the population ecology of open habitat species, like European peony (*Paeonia officinalis*), which is now very rare.

Andrieu et al. (2017) have explored the effect of forest spread on a few populations of the Massif Central (Southern France). They have estimated the Malthusian rate of increase including environmental stochasticity. For a site called TAU, the ecologists have estimated that the median finite annual growth rate is 1.025. They have also estimated the 95% confidence interval as $1.01 - 1.04$. After a logarithmic transformation, this interval corresponds to the average instantaneous growth rate $\pm$ 2 SD (standard deviations of the normal distribution).

(a) Estimate the variance σ_ϵ^2 of the environmental stochasticity (i.e., the variance of the logarithm of the multiplicative noise);

(b) assume that the peony abundance in 2018 in the site TAU is 211 individuals; calculate the median abundance in 2028;

(c) calculate the probability that the population falls below 310 individuals in 2036 (use a standard normal distribution table).

Problem ER16 The
European peony

Problem ER17

The Amur tiger (*Panthera tigris altaica*) is a flagship species of the boreal forest ecosystem in northeastern China and Russia Far East. During the past century, the tiger population has declined sharply.

Yu et al. (2011) have conducted a Population Viability Analysis (PVA) assuming that an initial population of 200 tigers is subject to environmental stochasticity for 100 years. They have estimated that, in the baseline scenario (no further deterioration of the tiger population), the population would grow in a Malthusian way and (1) the median population size at the end of 100 years would be 380 tigers, (2) the probability in 100 years for the population size to be lower than 200 would be 10%.

Based on the results of the PVA, estimate

(i) the median finite rate of increase of the Amur tiger,

Problem ER17 The Amur
tiger

(ii) the value of the environmental variance (that is the value σ_ϵ^2 of the variance of the logarithm of the population size) by using a table of the standard normal distribution.

Problem ER18

The population of brown bear (*Ursus arctos*) in Trentino-Italy has been steadily increasing (Rapporto orso e grandi carnivori 2016, https://grandicarnivori.provincia. tn.it). Here below you find the most recent statistics on the total number of bears

Year	2008	2009	2010	2011	2012	2013	2014	2015	2016
Numbers	20	26	27	33	29	40	37	38	38

Assume that *environmental stochasticity* only has been affecting the population in recent years. From the table estimate:

(I)   the instantaneous rate of demographic increase and the corresponding finite rate,
(II) the variance σ_ϵ^2 of the logarithm of the multiplicative noise,
(III) the probability that the population will fall below 50 individuals in 2026 (use a table of the normal distribution).

Problem ER18 The brown bear

From other data, the Rapporto orso (2016) estimates an average survival between subsequent years of 88%. Also, it reports that back in 2002 there were 8 bears (sex ratio 1:1). Suppose that in 2002 you wanted to estimate the long-term chance of extinction of the bear population subject to *demographic stochasticity* only. Use the demographic rates from the Rapporto orso (2016) to estimate the long-term probability of persistence of the brown bear in Trentino as resulting from the number of bears in 2002.

Problem ER19

The golden-beak parrots (*Parrotus chrysaetos*, an invented species) were captured in the past for ornamental purpose; in recent times they have been protected in the Montes Rojos reserve. Here below you can find the population numbers as recorded by the Guanavaca Ornithological Institute

Year	1992	1993	1994	1995	1996	1997	1998	1999	2000	2001	2002	2003
Numbers	21	98	285	379	273	313	290	345	330	288	301	314

Fit a Ricker model to the population data (*hint*: calculate the finite growth rates, then take logarithms and find the appropriate regression line). Find the nontrivial equilibrium of the deterministic model and assess its stability. Which is the long-term average of the parrot population at Montes Rojos?

Problem ER20

Dennis and Taper (1994) report the population numbers of the red deer (*Cervus elaphus*) population in the Yellowstone Park.

Year	1968	1969	1970	1971	1972	1973	1974	1975	1976	1977	1978	1979
Numbers	3172	4305	5543	7281	8215	9981	10529	12607	10807	10741	11855	10768

First, establish whether a Malthusian model can correctly describe the red deer dynamics. If the answer is negative, see whether a Ricker model can aptly describe the time evolution of the deer population. In any case find the variance of the environmental stochasticity σ_ε^2 (namely the variance of the logarithm of the multiplicative noise that influences the population abundance).

Problem ER20   A red deer
in the Yellowstone Park

Cervus elaphus

References

Andrieu E, Besnard A, Fréville H, Vaudey V, Gauthier P, Thompson JD, Debussche M (2017) Population dynamics of *Paeonia officinalis* in relation to forest closure: From model predictions to practical conservation management. *Biological Conservation* 215:51–60

Casas-Marce M, Soriano L, López-Bao JV, Godoy JA (2013) Genetics at the verge of extinction: insights from the Iberian lynx. *Molecular Ecology* 22:5503–5515

Dennis B, Taper ML (1994) Density dependence in time series observations of natural populations: Estimation and testing. *Ecological Monographs* 64:205–224

Furlan E, Stoklosa J, Griffiths J, Gust N, Ellis R, Huggins RM, Weeks AR (2012) Small population size and extremely low levels of genetic diversity in island populations of the platypus *Ornithorhynchus anatinus*. *Ecology and Evolution* 2:844–857

Hens H, Pakanen VM, Jäkäläniemi A, Tuomi J, Kvist L (2017) Low population viability in small endangered orchid populations: Genetic variation, seedling recruitment and stochasticity. *Biological Conservation* 210:174–183

Milne H, Robertson F (1965) Polymorphisms in egg albumen protein and behaviour in the eider duck. *Nature* 205:367–369

Ryman N, Baccus R, Reuterwall C, Smith MH (1981) Effective population size, generation interval, and potential loss of genetic variability in game species under different hunting regimes. *Oikos* 36:257–266

Schaub M, Zink R, Beissmann H, Sarrazin F, Arlettaz R (2009) When to end releases in reintroduction programmes: demographic rates and population viability analysis of bearded vultures in the alps. *Journal of Applied Ecology* 46:92–100

Velando A, Barros A, Moran P (2015) Heterozygosity-fitness correlations in a declining seabird population. *Molecular Ecology* 24:1007–1018

Wilkinson DM, O'Regan HJ (2003) Modelling differential extinctions to understand big cat distribution on Indonesian islands. *Global Ecology and Biogeography* 12(6):519–524

Yu T, Wu J, Smith A, Wang T, Kou X, Ge J (2011) Population viability of the siberian tiger in a changing landscape: Going, going and gone? *Ecological Modelling* 222:3166–3180

Part II
Populations in Spatially Explicit Landscapes

Chapter 5
Movement of Organisms and the Dynamics of Populations in Space

Space does matter at every scale of the hierarchical structure of biology, from nucleic acids to cells, from tissues to organisms, and from populations to the whole biosphere. Spatial patterns that we observe are clearly the result of aggregation phenomena that are constrained by fundamental biological or behavioural mechanisms. The shape of flowers, horns, shells, cones clearly reveals spatial organization, often characterized by a striking and fascinating regularity. Spatial patterning, though less regular and possibly time-varying, is also shown by many populations and ecosystems, e.g. schools of fish, flocks of birds or the tiger bush of many arid regions (see Fig. 5.1). Waves are also a typical spatial phenomenon that characterizes the functioning of many populations and ecosystems. Of particular importance to the present Anthropocene is the invasion of new species (often alien species that can contribute to the extinction of endemic ones) and the spread of pathogens.

The models illustrated so far are a very approximate description of simple ecological systems, because the real challenges that ecology poses to modelling come substantially from the complex spatio-temporal dynamics of populations, communities and ecosystems. Within this complexity, different scales are clearly seen, often hierarchically organized: for example, an exhaustive explanation of the functioning of a population of small mammals presupposes an understanding of (i) the physiology and behavior of individuals, which occurs on a short time scale (days), (ii) the vital cycle (growth, survival, reproduction), describable on a longer time scale (months), and (iii) the demography of the entire population, occurring on an even longer time scale (years). If we then think that every population is spatially organized and inserted in ecosystems whose essential components range from the microscopic dimensions of bacteria (with the relative space-time scales) to the macroscopic dimensions of large mammals or secular plants, we understand how ecology can be to modellers a challenging field that offers considerable opportunities for the application or development of new techniques. It is the integration of the different scales that poses the greatest challenges compared to mathematical tools that have traditionally been developed to explain relatively homogeneous physical and chemical systems. At the

Fig. 5.1   Aerial view of tiger bush in the W Bénin-Niger National Park. Average distance between two successive gaps in the vegetation is 50 m. Photo by Nicolas Barbier [CC BY-SA 3.0]

highest level of biological organization, the basic mechanism of spatial dynamics is the ability of organisms to move, leaving their location where they dwell for a time and colonizing new space. Even plants and fungi can actually move via vegetative reproduction (colonization of the surrounding space) or the release of propagules (seeds, spores). Understanding the implications of movement is therefore the fundamental approach to spatial ecology and is thus the subject of this chapter.

5.1   The Diffusion Process

The simplest way to describe the movement of organisms in space is to assume that it is random. If in addition one assumes that the population size N is large enough and makes some specific further hypotheses (which will be better defined below) one can get the so-called equation of transport and diffusion, which is the simplest model for the dispersal of a population of organisms in space (as well as for many other phenomena, such as the release of chemical substances in water and air). Let us see how to derive it by assuming for simplicity that space (indicated hereafter with x) is one-dimensional.

First, suppose that both space and time are discrete (Fig. 5.2). We denote by p the probability that an organism moves to the right with a step of length Δx in the time interval Δt and by $q = 1 - p$ the probability for an organism to move left. If we assume that in every spatial position there is a large number of individuals we can also say that p is the fraction of organisms that moves to the right and q is the fraction of organisms that moves to the left (according to the law of large numbers).

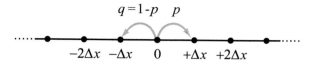

$$q = 1\text{-}p \quad p$$

$$-2\Delta x \quad -\Delta x \quad 0 \quad +\Delta x \quad +2\Delta x$$

Fig. 5.2 The scheme of random walk: an organism moves a step to the right with probability p or to the left with probability $1 - p$

Let us introduce the density (or concentration) of organisms in each position x at time t namely

$c(x, t)\Delta x$ = number of organisms that at time t are located between $x - \Delta x/2$ and $x + \Delta x/2$.

Then we can state that

$pc(x, t)\Delta x$ = Number of organisms that in the time interval Δt move from location x to location $x + \Delta x$;

$(1 - p)c(x, t)\Delta x$ = number of organisms that in the time interval Δt move from location x to location $x - \Delta x$.

We can then write the following balance equation

$$c(x, t + \Delta t)\Delta x = pc(x - \Delta x, t)\Delta x + (1 - p)c(x + \Delta x, t)\Delta x.$$

Divide now by Δx and develop the left- and right-hand side terms in Taylor series, thus getting

$$c(x,t) + \frac{\partial c}{\partial t}\Delta t + \mathcal{O}(\Delta t^2) = p\left[c(x,t) - \frac{\partial c}{\partial x}\Delta x + \frac{1}{2}\frac{\partial^2 c}{\partial x^2}\Delta x^2 + \mathcal{O}(\Delta x^3) \right] +$$

$$+ (1 - p)\left[c(x,t) + \frac{\partial c}{\partial x}\Delta x + \frac{1}{2}\frac{\partial^2 c}{\partial x^2}\Delta x^2 + \mathcal{O}(\Delta x^3) \right]$$

where with $\mathcal{O}(z)$ we indicate terms of order larger than or equal to z. If we simplify some of the terms left and right of the equal sign and divide by Δt we obtain

$$\frac{\partial c}{\partial t} + \frac{\mathcal{O}(\Delta t^2)}{\Delta t} = (1 - 2p)\frac{\partial c}{\partial x}\frac{\Delta x}{\Delta t} + \frac{1}{2}\frac{\partial^2 c}{\partial x^2}\frac{\Delta x^2}{\Delta t} + \frac{\mathcal{O}(\Delta x^3)}{\Delta t}.$$

We must now let the space step Δx and the time step Δt tend to zero to obtain the appropriate equation in continuous space and time. However, some additional assumptions are needed to obtain the so-called diffusion approximation. More precisely we have to assume that

$$\frac{\Delta x}{\Delta t} = \text{Absolute movement speed of a single organism} \to \infty$$

$$\frac{1}{2}\frac{\Delta x^2}{\Delta t} \to D \text{ Finite positive constant}$$

$$(2p - 1)\frac{\Delta x}{\Delta t} = p\frac{\Delta x}{\Delta t} + (1 - p)\frac{-\Delta x}{\Delta t} = \text{average population speed finite constant.}$$

In particular, note that the first and the third assumption imply that $2p - 1$ must necessarily tend to zero, or, stated otherwise, that the diffusive approximation is valid if the probability to move right or left differ only by a small amount, i.e. differ very little from $1/2$.

Letting Δx and Δt tend to zero, one obtains

$$\frac{\partial c}{\partial t} = -v\frac{\partial c}{\partial x} + D\frac{\partial^2 c}{\partial x^2} \tag{5.1}$$

which is the famous *transport (or advection or drift) and diffusion equation*. If v is zero one gets the equation of pure diffusion.

Parameter D is termed diffusion coefficient. Note that its measurement unit is a squared distance per unit time. Note also that the transport term is proportional to $\frac{\partial c}{\partial x}$, i.e. the concentration gradient. It must be remarked that the assumption that every organism is moving at very large speed (infinite in the limit) is crucial to obtaining Eq. 5.1. In fact, if we had supposed that the speed of each organism were finite, then Δx would tend to zero like Δt and then Δx^2 would tend to zero faster than Δt. As a result, the diffusive term would vanish and only advection would be operating. Using other assumptions leads to more realistic models than simple diffusion (an account is for instance provided in Rinaldo et al. 2020), but for many applications this simple model is quite effective.

It is easy to understand that the transport term in Eq. 5.1 has just a translation effect on the solution of the same equation. In other words, the solutions of Eq. 5.1 can be obtained from solutions of the pure diffusion equation

$$\frac{\partial c}{\partial t} = D\frac{\partial^2 c}{\partial x^2} \tag{5.2}$$

provided one replaces the space coordinate x with the so-called "moving reference frame coordinate", or $x + vt$. For this reason, we will from now on focus exclusively on Eq. 5.2.

A solution of Eq. 5.2 is completely determined if one specifies suitable initial conditions and suitable conditions at the space boundary. Providing initial conditions means to specify the organisms density in each location x at the initial instant, conventionally denoted by 0, namely

$$c(x, 0) = c_0(x) = \text{a given function of space.}$$

As regards the boundary conditions, there are several possible cases, depending on the spatial domain considered. In the next sections we will deal with the following two situations only.

Infinite Domain (Cauchy Problem)

If the domain is infinite, x can take any value between $-\infty$ and $+\infty$. From a practical standpoint, this condition is that of a population that can disperse without finding virtually any spatial barrier. Note that in this case there is no real boundary. As the only phenomenon that drives changes in the density of organisms at each point is just movement, the following important condition must be verified

$$\text{Total number of organisms} = \int_{-\infty}^{+\infty} c(x,t)\, dx = \int_{-\infty}^{+\infty} c_0(x)\, dx = N \text{constant}.$$

Therefore the solution $c(x,t)$ must be a bounded function at each time instant and such that

$$\lim_{x \to +\infty} c(x,t) = \lim_{x \to -\infty} c(x,t) = 0.$$

It is possible to prove that the solution of the Cauchy problem of Eq. 5.2 is unique if we consider bounded functions only.

Finite Domain (Dirichlet Problem)

The simplest finite domain in a one-dimensional space is a segment of length L, namely $x \in [0, L]$. As a practical reference case, you may imagine that the territory is an island or a wood which is very narrow, oblong and of length L (see Fig. 5.3a), and the habitats outside the island or the forest are totally unsuitable for the survival and movement of the dispersing organisms. Then the boundary conditions become

$$c(0, t) = c(L, t) = 0.$$

In mathematical terms this problem is often called *absorbing barrier problem*.

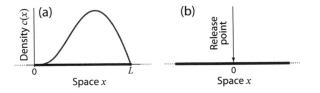

Fig. 5.3 a One-dimensional problem with absorbing barrier (Dirichlet problem); **b** A certain number of organisms is released in a point (conventionally located at $x = 0$) of a one-dimensional infinite space and randomly disperses without finding any barrier to its movement (Cauchy problem)

5.2   Diffusion in an Infinite Domain

Suppose that a large enough number N of animals is released at a point in space (which we conventionally indicate with $x = 0$) and that the organisms can disperse without any barrier to their diffusion (see Fig. 5.3b). First we note that we can introduce the function

$$p(x, t) = c(x, t)/N.$$

It represents the fraction of organisms per unit length that at time t is at distance x from the release site. More precisely

$p(x, t)dx = $ fraction of the N organisms that at time t are located between x and $x + dx$.

It should be noted that $p(x, t)$ has the same properties as a probability density, in particular the property

$$\int_{-\infty}^{+\infty} p(x, t)dx = 1$$

at any time t.

One can easily verify by direct substitution into Eq. 5.2 that the solution of the problem is nothing but a Gaussian distribution with respect to space, characterized by a time-increasing variance; more precisely

$$c(x, t) = Np(x, t)$$

with

$$p(x, t) = \frac{1}{\sqrt{4\pi Dt}} \exp\left(-\frac{1}{2}\frac{x^2}{2Dt}\right). \tag{5.3}$$

As time varies, the mean value—which is also the mode and the median—of $p(x, t)$ is always null because, without the transport term, organisms evenly spread to the right and to the left. Instead, the variance grows linearly with time, and hence the concentration profile of individuals becomes flatter and flatter (see Fig. 5.4). In

Fig. 5.4 Time behaviour of
the solution to the diffusion
problem when an initial
number of organisms is
released at $x = 0$. The graph
shows various snapshots of
the solution at different
instants

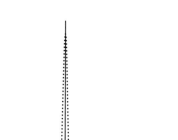

0 ——————————

0

Space x

particular, the standard deviation is given by

$$\sigma_x = \sqrt{2Dt}$$

and thus increases with the root of time.

It is to be noted that for $t = 0$ the variance vanishes, so the N organisms are all concentrated in the origin. Therefore the solution meets the initial conditions. In fact, since organisms are all simultaneously released at the same point, $c_0(x)$ is a pulse (technically a Dirac delta) located in $x = 0$. One can also compute the average absolute distance where organisms are located at time t after the release. With simple calculations one gets

$$E\left[|x|\right] = \int_{-\infty}^{+\infty} |x|\, p(x,t)dx = 2 \int_{0}^{+\infty} x p(x,t)dx = \sqrt{\frac{4Dt}{\pi}} \cong 1.128\sqrt{Dt}.$$

The diffusion equation is easily generalized to the case when we consider more than one spatial dimension. In particular when organisms disperse in a plane, described by the spatial coordinates x and y, and the diffusion coefficient is the same in all directions (*isotropic diffusion*), the diffusion equation is

$$\frac{\partial c}{\partial t} = D\left(\frac{\partial^2 c}{\partial x^2} + \frac{\partial^2 c}{\partial y^2}\right). \tag{5.4}$$

The solution to the problem of the release of N organisms in a point of a plane is just a bivariate normal distribution. More precisely, if we introduce the fraction of individuals per unit area $p(x, y, t)$ defined as

$p(x, y, t)dxdy$ = fraction of the N organisms that at time t are located in the square of area $dxdy$ whose lower left corner has coordinates (x, y)

the solution of Eq. 5.4 is given by

$$c(x, y, t) = Np(x, y, t)$$

with

$$p(x, y, t) = \frac{1}{4\pi Dt} \exp\left(-\frac{1}{2}\frac{x^2 + y^2}{2Dt}\right).$$

At each time instant t the fraction $p(x, y, t)$ has precisely the shape of a symmetrical bell (see Fig. 5.5a). Contour lines are therefore circles (Fig. 5.5b).

It is possible to prove (see the book by Pielou 1977) that the fraction f_R of the population that lies outside the contour line of radius R at time t is given by

$$f_R = \exp\left(-\frac{R^2}{4Dt}\right).$$

Therefore, the radius of the contour line that contains for example 99% of the population at a given instant t satisfies the equation

$$0.01 = \exp\left(-\frac{R^2}{4Dt}\right)$$

and thus is given by

$$R = 2\sqrt{\ln 100}\sqrt{Dt} = 4.292\sqrt{Dt}.$$

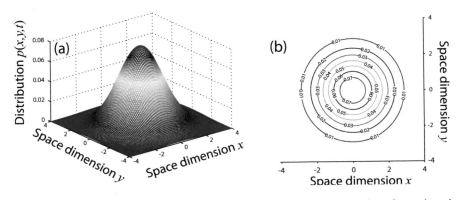

Fig. 5.5 Solution to the diffusion problem in the plane when an initial number of organisms is released at the point of coordinates $x = 0$, $y = 0$. (a) Shape of $p(x, y, t)$ with $D = 2$ and $t = 0.5$; (b) corresponding contour lines

It increases with the root of time, while the area that contains a certain fraction of the population (πR^2) grows linearly with time.

5.3 Diffusion in a Finite Habitat

In this case the goal is to find a solution to the Dirichlet problem with absorbing barrier, namely a solution of Eq. 5.2 that satisfies the following initial and boundary conditions

$$c(x, t) \geq 0 \qquad\qquad \text{for any } t \geq 0 \text{ and for } 0 \leq x \leq L$$
$$c(0, t) = c(L, t) = 0 \qquad \text{for any } t \geq 0$$
$$c(x, 0) = c_0(x) = \text{given function.}$$

Of course, even the initial density must satisfy the condition of absorbing barrier, i.e. $c_0(0) = c_0(L) = 0$ (see Fig. 5.6).

The way for solving the absorbing barrier problem requires the use of the Fourier series. According to the theory developed by Fourier, a continuous function $f(x)$ that is periodic of period P can be expanded as the sum of a constant and an infinite number of sinusoidal components of frequencies that are multiples of the fundamental frequency $1/P$. More explicitly, if we indicate the wave number with k, the frequency with k/P, and the angular frequency with $2\pi k/P$, it turns out that

$$f(x) = \frac{a_0}{2} + \sum_{k=1}^{\infty} a_k \cos\left(\frac{2\pi k}{P}x\right) + b_k \sin\left(\frac{2\pi k}{P}x\right) \tag{5.5}$$

$$a_k = \frac{2}{P} \int_0^P f(x) \cos\left(\frac{2\pi k}{P}x\right) dx = \frac{2}{P} \int_{-\frac{P}{2}}^{\frac{P}{2}} f(x) \cos\left(\frac{2\pi k}{P}x\right) dx \tag{5.6}$$

$$b_k = \frac{2}{P} \int_0^P f(x) \sin\left(\frac{2\pi k}{P}x\right) dx = \frac{2}{P} \int_{-\frac{P}{2}}^{\frac{P}{2}} f(x) \sin\left(\frac{2\pi k}{P}x\right) dx \tag{5.7}$$

$$\frac{a_0}{2} = \frac{1}{P} \int_0^P f(x) dx = \text{mean value.} \tag{5.8}$$

The Fourier series is quite useful for our purposes because it allows us to use the following trick. The function $c_0(x)$ defined in the interval $0 \leq x \leq L$, can be seen as just a part of a periodic function of period $2L$ defined on the whole real axis (see Fig. 5.7). This latter can then be expanded in Fourier series

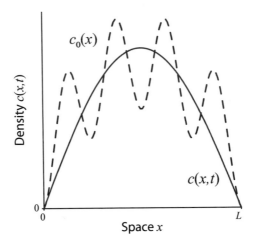

Fig. 5.6 Diffusion problem in a finite one-dimensional domain with absorbing barrier

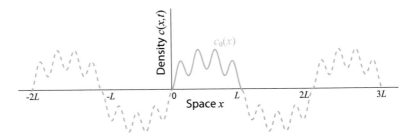

Fig. 5.7 The solution to the absorbing barrier problem can be seen as part of a periodic function defined on the whole real axis

$$\frac{a_0}{2} + \sum_{k=1}^{\infty} a_k \cos\left(\frac{2\pi k}{2L}x\right) + b_k \sin\left(\frac{2\pi k}{2L}x\right).$$

Moreover, the periodic function of which $c_0(x)$ is just a part has zero mean value and vanishes at $x = 0$ and $x = L$. Thus, according to the properties of Fourier series, the coefficients of the series must necessarily be such that $a_0 = 0$ and $a_k = 0$ for any $k \geq 1$. In other words, we can state

$$c_0(x) = \sum_{k=1}^{\infty} b_k \sin\left(\frac{\pi k}{L}x\right)$$

namely think of the function $c_0(x)$ as the weighted sum (with weights equal to b_k) of sine waves with frequencies equal to $1/2L, 2/2L = 1/L, 3/2L, 4/2L = 2/L$ and so on. The sinusoidal components are also called modes or modal components and k is called wave number. The properties of $c_0(x)$ are actually shared by all the functions

$c(x, t)$ with $t > 0$, because they too have to satisfy the absorbing barrier condition. Therefore, these functions can be expanded into the following Fourier series

$$c(x, t) = \sum_{k=1}^{\infty} B_k(t) \sin\left(\frac{\pi k}{L} x\right). \tag{5.9}$$

The coefficients $B_k(t)$ of the Fourier series in Eq. 5.9 must be such that the function $c(x, t)$ is a solution to the diffusion equation. First of all, we must necessarily have

$$c(x, 0) = c_0(x)$$

which implies $B_k(0) = b_k$. Secondly, we require that

$$\frac{\partial c}{\partial t} = D \frac{\partial^2 c}{\partial x^2}.$$

Since

$$\frac{\partial c}{\partial t} = \sum_{k=1}^{\infty} \frac{d B_k}{dt} \sin\left(\frac{\pi k}{L} x\right)$$

$$\frac{\partial c}{\partial x} = \sum_{k=1}^{\infty} B_k(t) \frac{\pi k}{L} \cos\left(\frac{\pi k}{L} x\right)$$

$$\frac{\partial^2 c}{\partial x^2} = -\sum_{k=1}^{\infty} B_k(t) \frac{\pi^2 k^2}{L^2} \sin\left(\frac{\pi k}{L} x\right)$$

it turns out that

$$\frac{d B_k}{dt} = -D \frac{\pi^2 k^2}{L^2} B_k \quad \text{with } k = 1, 2, \dots \tag{5.10}$$

The solution of Eq. 5.10 is

$$B_k(t) = B_k(0) \exp\left(-D \frac{\pi^2 k^2}{L^2} t\right)$$

and therefore the final formula for the weights of modal components of $c(x, t)$ is:

$$B_k(t) = b_k \exp\left(-D \frac{\pi^2 k^2}{L^2} t\right). \tag{5.11}$$

From this formula we conclude that all the modes fade out exponentially and hence that $c(x, t) \to 0$ for $t \to \infty$. This second result is obvious because the random movements slowly lead all the organisms out of the habitat suitable for the species. However, we note that the damping exponent of each mode increases with

Fig. 5.8 Time evolution of the solution to the diffusion problem with an absorbing barrier. The plot shows various snapshots at successive times, with the initial density being given by the red solid line. The total number of organisms in the domain is given by the area under each curve and decreases with time

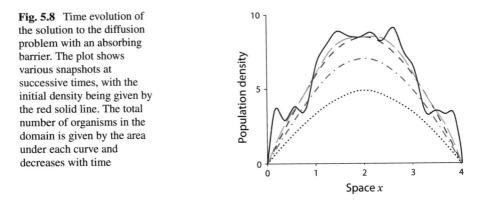

the square of the wave number k. Therefore, the average extinction time of the modal components with larger spatial frequency is much smaller. In other words, diffusion tends to quickly eliminate any initial spatial wiggling of organisms' density, as shown in Fig. 5.8 which reports different snapshots of $c(x, t)$ at successive times. The solid line indicates the initial density $c_0(x)$.

5.4   Demographic Increase and Diffusion

From the ecological viewpoint things get much more interesting when, in addition to movement, we introduce the population demography. In other words, we consider that population density at each spatial location changes over time not only because of movement, but also because of occurring births and deaths. Suppose that the birthrate v and mortality rate μ that operate in a certain spatial location depend only upon local density $c(x, t)$. If we indicate the per capita rate of increase by R we can write

$$R(c) = v(c) - \mu(c).$$

Then the time variation of density at each location x is given by the following equation

$$\frac{\partial c}{\partial t} = D \frac{\partial^2 c}{\partial x^2} + R(c)c. \tag{5.12}$$

Equation 5.12 is called *reaction-diffusion equation* because this same equation is used to describe the kinetics of a chemical reactor in which R is the speed at which a chemical component with concentration c is formed. The analysis of the reaction-diffusion equation is not so simple in the general case of any $R(c)$, but is quite simple when the demography is Malthusian, namely when the per capita growth rate R is independent of density:

$$R(c) = r \text{ constant.}$$

In fact, Eq. 5.12 reduces to

$$\frac{\partial c}{\partial t} = D\frac{\partial^2 c}{\partial x^2} + rc$$

and can be easily solved via a change of variables. By introducing the new variable

$$z(x, t) = \exp(-rt)c(x, t)$$

it is easy to derive that

$$\frac{\partial z}{\partial t} = -r\exp(-rt)c(x, t) + \exp(-rt)\frac{\partial c}{\partial t} = -rz + \exp(-rt)\left[D\frac{\partial^2 c}{\partial x^2} + rc\right] = D\frac{\partial^2 z}{\partial x^2}.$$

Therefore, z satisfies a pure diffusion equation whose solutions we have already learnt how to derive in the previous section. Let us separately consider the two Cauchy and Dirichlet problems previously defined.

5.4.1 Malthusian Growth and Dispersal in an Infinite Habitat

Suppose we introduce a new species or reintroduce one that was once present in the environment by releasing a number N_0 of individuals, in the appropriate sex ratio, at time $t = 0$ at location $x = 0$. These organisms will disperse without barriers in space, will reproduce and will die. How docs the organisms' density vary in locations far away from that of release? If $r > 0$, it is interesting to wonder whether the density at every location is going to increase or whether dispersal prevents the local growth of the population. From Eq. 5.3 we know that

$$z(x, t) = N_0 \frac{1}{\sqrt{4\pi Dt}} \exp\left(-\frac{1}{2}\frac{x^2}{2Dt}\right)$$

and thus we easily derive that

$$c(x, t) = N_0 \frac{1}{\sqrt{4\pi Dt}} \exp\left(rt - \frac{1}{2}\frac{x^2}{2Dt}\right). \tag{5.13}$$

Figure 5.9 shows the evolution of density in successive time instants. One can remark for example that at the release point $x = 0$ the density initially decreases because diffusion is prevailing, but in the long term population growth prevails because the term $\exp(rt)$ has the most influence on the population evolution for large t. In spatial sites that were initially unpopulated, density is always growing: the locations are reached via diffusion and then basically population keeps growing because of demography. It is obvious that the total number of organisms grows exponentially because

Fig. 5.9 Time evolution of the solution to the problem of Malthusian growth and diffusion in an unbounded habitat. The figure shows various snapshots of the solution in successive instants

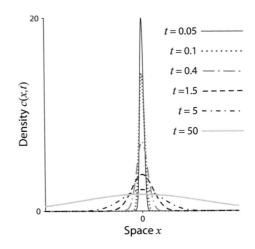

$$N(t) = \int\limits_{-\infty}^{+\infty} c(x,t)\,dx = \exp(rt) \int\limits_{-\infty}^{+\infty} z(x,t)\,dx = N_0 \exp(rt)$$

If we consider the problem from a more realistic viewpoint, i.e. the organisms are released and can move in a two-dimensional environment, the solution is simply given by

$$c(x, y, t) = N_0 \frac{1}{4\pi Dt} \exp\left(rt - \frac{1}{2}\frac{x^2 + y^2}{2Dt}\right).$$

From this formula we can gather how the population colonizes new space. It is possible to prove (Pielou 1977) that the number N_{out} of individuals in the population that at time t are outside a contour line of radius R_t is given by

$$N_{out} = N_0 \exp\left(rt - \frac{R_t^2}{4Dt}\right).$$

It is reasonable to assume that the presence of individuals is no longer detectable when N_{out} is below a certain threshold fraction f_{min} (for example 1%) of the number N_0 of initially released organisms. Therefore the radius within which the entire population is in practice contained satisfies the equation

$$f_{min} = \exp\left(rt - \frac{1}{2}\frac{R_t^2}{2Dt}\right)$$

and thus

$$R_t^2 = 4Drt^2 - 4Dt \ln f_{min}. \tag{5.14}$$

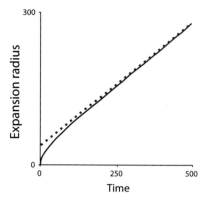

Fig. 5.10 Expansion radius as a function of time for a Malthusian population dispersing via diffusion (blue curve). f_{min} is set to 1%. Asymptotically the radius increases linearly. The dashed red straight line is the asymptote which is given by $2\sqrt{Dr}t - \frac{D\ln(f_{min})}{\sqrt{Dr}}$

With the passing of time the first term of Eq. 5.14 becomes much larger than the second, hence we can conclude that, if we include Malthusian growth in addition to diffusion, the radius marking population expansion, after an initial transient, grows approximately linearly with time, not with the root of the time (see Fig. 5.10). The colonization speed is thus approximately equal to $2\sqrt{Dr}$. Equivalently, we can say that the area which virtually contains all the population increases with the square of time.

There are several examples that confirm the validity of this simple model. Many refer to the introduction of alien species. The classical example is that of the muskrat (see Fig. 5.11), a rodent native to North America that was imported to Bohemia at the beginning of 1900's to exploit its fur. It looks like five animals managed to escape from a breeding farm near Prague in 1905. The muskrat population in the wild began then to increase colonizing space at a speed of about 12 km per year. Another example, that of the Argentine ant *Linepithema humile* (at that time named *Iridomyrmex humilis*) introduced in some areas of western United States, is summarized in Fig. 5.12. Table 5.1 shows some values of expansion velocity as observed in both terrestrial and marine invasive species.

5.4.2 Malthusian Growth and Diffusion in a Finite Habitat with an Absorbing Barrier

Suppose that in a wood, suited to the ecological requirements of an insect, there are initially N_0 individuals distributed with density $c_0(x)$ (for example as in Fig. 5.3a). What will happen in the long run? Will diffusion prevail, which tends to bring insects out of the wood, or population growth, which tends to let population increase inside

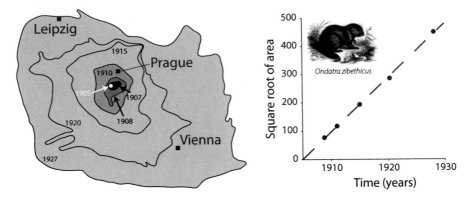

Fig. 5.11 The expansion of muskrat (*Ondatra zibethicus*, antique animal illustration as inset) in Europe following the accidental release of five muskrats in the location shown as a white dot in panel **a**, near Prague. Maps are redrawn after the original reported in Elton (1958). **b** If we approximate the areas (km^2) occupied by the muskrats with circles, their root increases linearly with time, as reported in the seminal paper by Skellam (1951). By dividing the slope of the fitting line by $\sqrt{\pi}$ one obtains the average radius increase as kilometres per year. The estimated value is 11.8 km/year

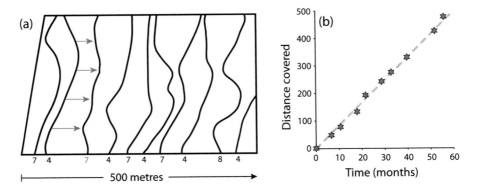

Fig. 5.12 The expansion of Argentine ant *Linepithema humile* in a meadow close to San Diego (California), initially occupied by the Californian ant *Pogonomyrmex californicus*. The experiment took place between October 1963 and October 1968. **a** Solid lines represent the invasion wave fronts of the Argentine ant (left to right). Elapsed times (months) between one front and the next are indicated below each solid line. **b** Distance (metres) from release location averaged along the vertical transect as a function of time. Redrawn after Fig. 1 by Erickson (1971)

the wood? From Eqs. 5.9 and 5.11 we know that

$$z(x, t) = \sum_{k=1}^{\infty} b_k \exp\left(-D\frac{\pi^2 k^2}{L^2} t\right) \sin\left(\frac{\pi k}{L} x\right)$$

where b_k are the weights of the modal components of $c_0(x)$. Then we easily derive that the solution to the problem of growth and diffusion in a finite interval of length

Table 5.1 Various expansion speeds for a few terrestrial and marine species (Grosholz 1996)

Species name (common name)	Observed speed (km/year)
Terrestrial species	
Impatiens glandulifera (Himalayan balsam)	9.4–32.9
Lymantria dispar (Asian gypsy moth)	9.6
Pieris rapae (small white butterfly)	14.7–170
Oulema melanopus (cereal leaf beetle)	26.5–89.5
Ondatra zibethicus (muskrat)	0.9–25.4
Sciurus carolinensis (grey squirrel)	7.66
Streptopelia decaocto (collared dove)	43.7
Sturnus vulgaris (European starling)	200
Yersinia pestis (Bubonic plague bacterium)	400
Marine species	
Botrylloides leachi (colonial tunicate)	16
Membranipora membranacea (lacy crust bryozoan)	20
Carcinus maenas (green crab)	55
Hemigraspus sanguineus (Asian shore crab)	12
Elminius modestus (acorn barnacle)	30
Littorina littorea (common periwinkle)	34
Mytilus galloprovincialis (Mediterranean mussel)	115
Perna perna (brown mussel)	95

L with an absorbing barrier at the interval extremes is given by

$$c(x, t) = \exp(rt)z(x, t) = \sum_{k=1}^{\infty} b_k \exp\left[\left(r - D\frac{\pi^2 k^2}{L^2}\right)t\right] \sin\left(\frac{\pi k}{L}x\right). \quad (5.15)$$

This relationship is extremely important because it tells us that $c(x, t)$ is the sum of modal components that might dampen or amplify over time according to the exponential coefficient being positive or negative, that is depending on

$$r - D\frac{\pi^2 k^2}{L^2} > 0 \quad \text{or} \quad r - D\frac{\pi^2 k^2}{L^2} < 0.$$

The crucial remark is that the coefficient $r - D\frac{\pi^2 k^2}{L^2}$ is a decreasing function of k^2 and thus the mode with the highest growth rate over time is the fundamental mode (i.e. with wave number $k = 1$). Therefore if

$$r - D\frac{\pi^2}{L^2} < 0$$

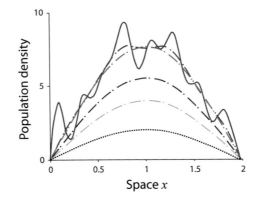

Fig. 5.13 Behaviour of the solution to the problem of Malthusian growth and diffusion in a finite one-dimensional habitat with a length that is smaller than the critical size. The figure shows various snapshots of the solution in successive time instants. Parameters are $D = 1$, $r = 2$, $L = 2$. The critical habitat length is 2.221. The solid line displays the initial density profile $c_0(x)$

namely if

$$L < \pi\sqrt{\frac{D}{r}} = L_{cr} \qquad (5.16)$$

then all the modal components will vanish in the long time and the function $c(x, t) \to$ 0 for $t \to \infty$. Equation 5.16 is one of the most important results of reaction-diffusion theory. It tells us that if the size of the habitat suitable for a species is smaller than a critical value $L_{cr} = \pi\sqrt{\frac{D}{r}}$ then the population is surely doomed to extinction. It is worthwhile to note the fundamental difference with the case of growth and diffusion in a virtually unbounded domain: in that case the only condition for population extinction was that the per capita growth rate r be negative, whereas here extinction is possible even with $r > 0$ if the habitat size is smaller than the critical threshold.

In addition we note that the damping coefficient of each mode increases with the square of the wave number k. Therefore, the modal components of higher frequency will dampen much more rapidly (see example in Fig. 5.13).

On the contrary, if the habitat size L is larger than L_{cr}, the damping regards only the modal components with wave number k large enough, precisely those for which

$$k > \frac{L}{\pi}\sqrt{\frac{r}{D}}.$$

Instead, the low-frequency modes with wave number such that $r - D\frac{\pi^2 k^2}{L^2} > 0$ grow exponentially. The fundamental mode ($k = 1$, period $= 2L$) is the one growing most rapidly with a rate of increase given by

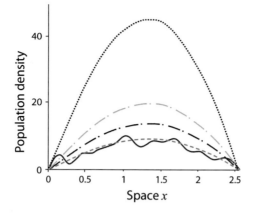

Fig. 5.14 Behaviour of the solution to the problem of Malthusian growth and diffusion in a finite one-dimensional habitat with a length that is larger than the critical size. The figure shows various snapshots of the solution in successive time instants. Parameters are $D = 1, r = 2, L = 2.6$. The critical habitat length is 2.221. The fundamental mode is the only one with a positive rate of increase. The solid line displays the initial density profile $c_0(x)$

$$r - D\frac{\pi^2}{L^2}.$$

Thus we can conclude that if $L > L_{cr}$ population growth prevails over dispersal and the population can grow even within a confined habitat. Figure 5.14 shows an example of the solution behaviour in this case.

One can easily extend the analysis to the more realistic case in which diffusion occurs on a surface; then the diffusion equation in two dimensions must be employed. Of course, in this case the bounded habitat where the population is located can have different shapes: the two simplest are square and circular. In the first case (square with side L) the condition of absorbing barrier is

$$c(0, y, t) = c(L, y, t) = 0$$
$$c(x, 0, t) = c(x, L, t) = 0$$

while in the second case the condition is that $c(x, y, t) = 0$ for all the points satisfying the relationships $x^2 + y^2 = R^2$ (circle of radius R).

In a way quite similar to the one-dimensional case one can show that the population is doomed to extinction if the suitable habitat area A is smaller than a critical value A_{cr}. In particular, for a square habitat the critical size is

$$A_{cr} = 2\pi^2 \frac{D}{r} \tag{5.17}$$

and for a circular habitat it is

$$A_{cr} \approx 1.84\pi^2 \frac{D}{r}. \tag{5.18}$$

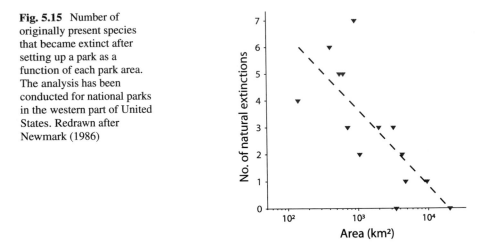

Fig. 5.15 Number of originally present species that became extinct after setting up a park as a function of each park area. The analysis has been conducted for national parks in the western part of United States. Redrawn after Newmark (1986)

Suppose you want to rescue a species threatened with extinction by placing a number of individuals within a reserve. Formulas 5.17 and 5.18 are fundamental to decide the size and shape of the reserve. Of course, to determine the critical area it is necessary to have at least rough estimates of the per capita growth rate and the diffusion coefficient.

Figure 5.15 is an indirect confirmation of the theory of critical habitat size: William Newmark (1986) analyzed several national parks in the West of United States and found that the species extinction probability significantly decreases with the park size.

5.4.3 Diffusion and Density-Dependent Population Growth

The Malthusian demographic model is not very realistic if the intrinsic rate of population increase r is positive. In fact, if we wait for a sufficiently long time the organisms' density in any point can become very large. Instead, we know that intraspecific competition acts to limit the growth rate for high densities and even makes it vanish at the carrying capacity.

The mathematical treatment of the reaction-diffusion equation in the case of non-Malthusian demographics is a bit complicated and will not be illustrated in detail. We just report the main results, which anyway are quite intuitive in the light of what we learned for the Malthusian case. The fundamental assumption that we will make is that the growth rate $R(c)$ is a decreasing function of density c, thus excluding the case of Allee effect (or depensation, see Sect. 2.3), which implies more complicated phenomena. For example demographics may be logistic

$$R(c) = r \left(1 - \frac{c}{K}\right),$$

with K being the carrying capacity.

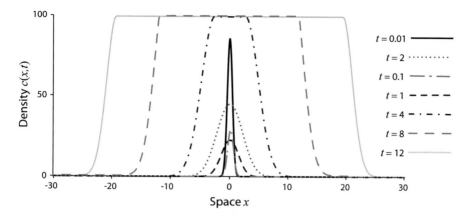

Fig. 5.16 Behaviour of the solution to the diffusion-reaction equation with density-dependent demography in an unbounded one-dimensional environment. The figure displays the snapshots of $c(x, t)$ in successive time instants. The carrying capacity K in any site is 100

Let us first analyse what happens in the case of an environment without barriers. For the Malthusian demographics described in the previous section, the density at any point tends to exponentially increase with time. If there is logistic dependence, instead, density in the long run tends to the carrying capacity K at any point in space. Therefore, if organisms are released in a given location (with coordinate $x = 0$), the solution behaviour is the one shown in Fig. 5.16. Initially, diffusion prevails, but then the demographic growth leads to saturation the zone close to the release site, and from that moment on there substantially occurs propagation of two wave fronts: one to the right and one to the left. One can show that the front speed is $2\sqrt{DR(0)}$. Note that this formula is the exact analogue of the asymptotic expansion velocity for the radius of the area containing a Malthusian population: in fact that radius was shown to grow as $2\sqrt{Drt}$. We should not wonder that the velocity of the front waves is influenced only by the value of $R(0)$ in the density-dependent growth rate, because the density in close proximity to the wave front in the direction of propagation is indeed very low, close to zero.

Even the solution of the case with finite habitat has a close analogy to what we learned for the Malthusian demographics. In fact, even with density dependence there exists a critical habitat size under which the population is doomed to extinction, because dispersal through the boundary prevails over population growth. To obtain expressions that provide the critical sizes it is sufficient to replace the Malthusian growth rate r with $R(0)$. Therefore, we obtain

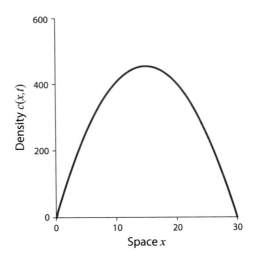

Fig. 5.17 Time behaviour of the solution to the reaction-diffusion equation with logistic demography in a finite one-dimensional domain ($L = 30$). The figure shows the stationary profile of density that is reached in the long run. The carrying capacity is 450 and is reached only in the center of the domain while in the other locations diffusion through the boundary hinders the attainment of carrying capacity. Density obviously vanishes at the boundary

$$L_{cr} = \pi\sqrt{D/R(0)} \qquad \text{for one-dimensional habitat}$$

$$A_{cr} = 2\pi^2 \frac{D}{R(0)} \qquad \text{for square habitat}$$

$$A_{cr} \approx 1.84\pi^2 \frac{D}{R(0)} \qquad \text{for circular habitat.}$$

Because of density-dependence $c(x, t)$ tends over the course of time towards a stationary profile of symmetrical shape, like that shown in Fig. 5.17.

References

Elton CS (1958) *The Ecology of Invasions by Plants and Animals.* Methuen & Co

Erickson JM (1971) The displacement of native ant species by the introduced Argentine ant Iridomyrmex humilis Mayr. Psyche 78:257–266

Grosholz ED (1996) Contrasting rates of spread for introduced species in terrestrial and marine systems. Ecology 77:1680–1686

Newmark W (1986) Species-area relationship and its determinants for mammals in Western North American national parks. Biological Journal of the Linnean Society 28:83–98

Pielou E (1977) Mathematical Ecology. Wiley, New York, NY

Rinaldo A, Gatto M, Rodriguez-Iturbe I (2020) River Networks as Ecological Corridors. Cambridge University Press, New York, Species, Populations, Pathogens

Skellam JG (1951) Random dispersal in theoretical populations. Biometrika 38:196–218

Chapter 6
Habitat Fragmentation and Destruction: The Dynamics of Metapopulations

6.1 Metapopulations and Habitat Loss

In the previous chapter we examined how the movement of organisms affects their spatial distribution and their demographics. However, the landscape in which they were able to disperse was always considered as uniform, though possibly bounded. But the hypothesis of homogeneity is very often unrealistic for two reasons.

The first—and increasingly dramatic—is that human activities are negatively affecting even the wildest and most pristine regions of the planet. One the most important consequences is fragmentation, that is the phenomenon of species habitat being divided in *fragments* (or *patches*, as they are frequently termed), which are immersed in a *matrix* of inhospitable territory. Perhaps, the most impressive testimony to the large spatial scale at which habitat destruction perpetrated by man operates is the fragmentation of the Amazon forest, the biggest green lung of our planet. It suffices to use an Internet browser, connect to one of the many sites providing Earth maps obtained via remote sensing and enter the string "Rondônia, Brazil" to directly ascertain what has been occurring in the recent past. The 400 km long and 100 km wide area that connects *Alto Paraíso* to *Primavera de Rondônia* is organized according to a geometric configuration that is so regular that there is no doubt about being the result of human intervention. In fact, an array of concurring circumstances, combined with a prolonged growing season and the low cost of land and labour, has indeed favoured the rapid conversion of entire regions of the Brazilian rain forest into soy-bean fields (Fearnside 2001). The collapse of the Peruvian *anchoveta* (*Engraulis ringens*) fishery, see also Sect. 8.2, has encouraged the use of soy in animal diet in both Europe and the United States, while an unexpected frost event in southern Brazil in 1975 has favoured the abandonment of coffee cultivation. The impact in terms of land fragmentation is not limited to the fields where soy-bean is planted. The construction of eight industrial waterways, three railways and many highways, all funded by the government to facilitate the production, transport and export of soy-bean has produced an inevitable avalanche effect, *efeito de arraste* in Brazilian. There have been many private investments (which have doubled the

© The Author(s), under exclusive license to Springer Nature Switzerland AG 2022
M. Gatto and R. Casagrandi, *Ecosystem Conservation and Management*,
https://doi.org/10.1007/978-3-031-09480-4_6

already substantial public investment) into felling trees or breeding cattle, additional elements of destruction and fragmentation. The Rondônia area unfortunately is not the only region of South America being subjected to such a destruction, as demonstrated, for example, by the satellite photos of the region around "San Pedro Limón, Bolivia". Not always the destruction of habitats is detectable by remote satellite sensing, because land-use change can still leave areas green, but devoid of the species that are key to the conservation of biodiversity in a given ecosystem. For example, looking at the region of the "Gunung Palung National Park, Kalimantan, Borneo", one still perceives those areas as being characterized by contiguous and uniform vegetation. However, the exploitation of the precious timber provided by the trees of the family *Dipterocarpaceae* in the period 1985–2001 caused the destruction of 56% of the Indonesian Borneo lowland forests (Curran et al. 2004). The plant species that now cover these areas are thus different from the original ones. If one considers that the *Dipterocarpaceae* are fundamental for 60–80% of birds and mammals living in that biome, the consequences for the protection of both plant and animal diversity are certainly dramatic. What is more dismaying is not only the progressiveness but also the rapidity at which destruction occurs. Even if Kalimantan is a protected area, the rate of deforestation in the park has increased by 9.5% annually from 1999 to 2004. Unprotected areas have even higher rates. And the destruction does not hit forests only, and not the tropics only.

A second reason to abandon the hypothesis of uniform landscape is that even territories without significant anthropic influence that may appear homogeneous at a certain spatial scale or for a certain species may not be so at other scales or for other species. For example, many butterflies are strict specialists—every mother chooses with meticulous care the plant on which to lay her eggs, the only one that can actually serve as food for the offspring (caterpillars). The females of the species *Euphydryas editha bayensis*, for example, emerge as adults in spring and lay their eggs almost exclusively on annual plants of the species *Plantago erecta*. These grow mostly on serpentine soils which have very special chemical characteristics (low Mg/Ca ratio and high concentration of heavy metals). Not only the distribution of these soils is very fragmented in Northern California, as shown in Fig. 6.1, but both the spatial pattern and the area of the different fragments is quite varied and uneven.

In particular, the figure shows that there exists a fragment of quite big size (*Morgan Hill*), where the butterfly population has always been present and plentiful. In other smaller fragments, instead, several extinctions of the butterflies have been recorded. We might call these extinctions local in both space and time, because at a later date some of the fragments that became empty were actually colonized again, as shown in the figure. This kind of dynamics in space and time regarding species spread over a fragmented landscape and subject to frequent local extinctions and colonizations is termed metapopulation dynamics in the ecological literature. Ehrlich and his colleagues have studied the dynamics of this natural metapopulation for about thirty years, starting from the 1960s (Ehrlich 1965). In more recent years, it has been noted that the geographic range of the species is moving towards North (about 2° on average) in response to climate change (Parmesan 1996).

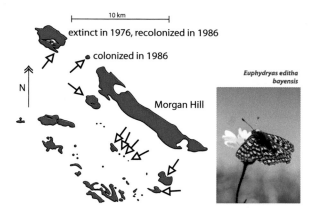

Fig. 6.1 Distribution of serpentine grasslands in Santa Clara County, California. White arrows indicate the locations of nine small populations of the butterfly *Euphydryas editha bayensis* (inset) measured via mark-recapture techniques in 1987. These local populations, with the Morgan Hill population in addition, constitute the metapopulation studied by Harrison et al. (1988)

The concept of metapopulation was proposed by Richard Levins (1969) who introduced a model that we will present later and has become a paradigm for conservation ecology. Curiously enough, this model was intended to identify control policies for the eradication of agricultural pest. As is known, most of the pests can spread on a large scale into a territory covered by crop fields thanks to their autonomous motion capacity, via both active movement (as some insects do through their flight) and passively (such as spores transported by the wind). In this regard, Levins remarked that the agricultural area on which to intervene in order to control an infestation is in general more extended than the single population of parasites (namely the single infested field or even the single infested plant). Thus Levins wrote "the control strategy must be defined for a *population of populations* in which local extinctions are balanced by emigration from other populations".

As suggested by the etymology of the word, the term meta-population refers to a population whose constituent parts are not individuals, but other populations which are called local (or sub-populations). We can say that while the term *local population* refers to organisms that inhabit an island of favourable habitat immersed in inhospitable surroundings, with *metapopulation* we mean the residents of the entire archipelago of habitats. Therefore there exist two spatial scales at which the dynamics of a metapopulation must be described and studied. One is related to the demographic events—birth and death processes—which take place exclusively at the level of local populations. The other spatial scale concerns the events of dispersal of individuals that move (or are moved by external vectors) between the different fragments and is thus termed global scale. The case of insect pests is particularly appropriate for emphasizing the difference between the two spatial scales. We could in fact say that there is *extinction* (*tout court*, without further specification) when all occurrences of local pests have been eradicated. Conversely, if even a single individual in a single local population does survive, then we say that there is *persistence* of the metapopulation. In the case in which the fragments are of small size compared to the needs of the species being analysed, local extinction can be quite frequent. Therefore, the likelihood for the metapopulation to escape extinction is highly dependent on the species ability to disperse and colonize empty fragments.

Although the model was originally introduced by Levins with the aim of under-standing how to destroy an unwanted population, its importance to conservationists is clear for two reasons. First of all, as we shall see below, it provides insight into the ecological conditions under which a population living in a fragmented environment is doomed to extinction. Second, this model has a very general validity because it allows the application of the island paradigm inherent to the metapopulation concept not only to the case of terrestrial pests, but to many other situations of clear ecologi-cal interest, which may seem to be non fragmented while they actually are. Quoting from another article by the same Levins (1970), "...the distribution of many species even on the mainland is insular. Mountain tops, lakes, individual host plants, a fallen log, a patch of vegetation, a mammalian gut, or, less obviously, a region of optimal temperature or humidity are all islands for the appropriate organisms. Therefore the insular model is much more broadly applicable."

6.2   Different Modelling Approaches to the Study of Metapopulations

Because of the concurrence of two different spatial scales, describing the dynamics of a metapopulation in quantitative terms is far from being a simple task. To clarify the problems arising in the choice of a suitable modelling approach an example can be usefully provided.

The amphibian *Rana muscosa* is an endemic species of the Sierra Nevada (USA). Once, it was very common in lakes, streams and ponds of mountain areas. For a variety of co-occurring reasons, still under scientific investigation, in the past cen-tury the species has disappeared from about 90% of the sites it previously occupied in California and Nevada. One of the identified causes is the predation upon tad-poles by non-native species of trout of the genera *Oncorhynchus spp.*, *Salmo spp.* and *Salvelinus spp.* which have been artificially introduced in mountain lakes since the middle of the past century to facilitate sport fishing (Knapp et al. 2007). Exper-imental removal of non-native trout (Vredenburg 2004) have demonstrated that the abundance of tadpoles and frogs goes back rather rapidly to values statistically indis-tinguishable from those of adjacent lakes where the fish have always been absent. Vance Vredenburg studied the effect of fish removal on the metapopulation structure of *R. muscosa* for a system of 81 mountain lakes in Sixty Lake Basin, Kings Canyon National Park, California. Figure 6.2 shows a simplified distribution map of frogs and trout in the major lakes of the basin. The five lakes indicated by a number are those where trout have been progressively removed over the five years from 1997 to

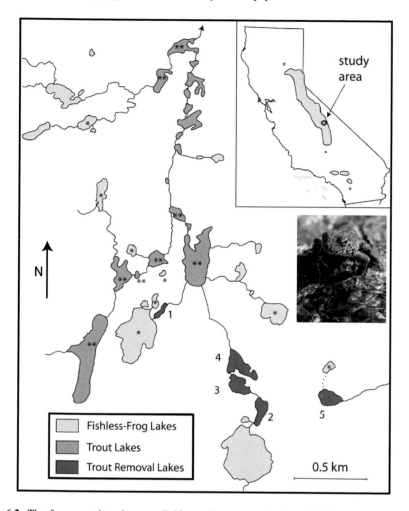

Fig. 6.2 The fragmented territory available to the metapopulation of California yellow-legged frogs (inset) in a system of mountain lakes in the Kings Canyon National Park (California, USA) as described by Vredenburg (2004). Copyright (2004) National Academy of Sciences, U.S.A

2001. Two more aspects deserve mentioning. First, in the year following the trout removal, each of the five lakes had a number of larvae and post-metamorphic frogs significantly different from that of lakes where fish were not removed. This is a signal that the trophic interaction played an important role in determining the disappearance of frogs. In the absence of predators, in fact, the density of both tadpoles and adult frogs have been rapidly increasing over the years, and have reached, at the end of the experiment, values of the order of 2–3 post-metamorphic frogs per linear decameter of coast. Second, the lake that responded more slowly to ecosystem restoration was lake number 5. Among the possible causes, Vredenburg includes the fact that this lake is not connected by rivers to other lakes where *R. muscosa* was present before starting the experiment. By looking again at the territorial distribution of lakes in Fig. 6.2 one can notice that while lake 1, as well as the lake subsystem 2–3–4, is downstream of a lake already occupied by frogs, lake 5 is upstream of any other lake of the river network; therefore it can be colonized more difficultly. In all probability, indeed, the colonization of the lake took place because of the migration of frogs present in the little grey lake next to lake 5. Although the little lake is not connected to lake 5, successful migration of *R. muscosa* individuals can evidently occur by other routes (for example, terrestrial). That is why Vredenburg has connected lake 5 with the small northern grey lake by a dashed arc.

Suppose we want to describe the dynamics of the species metapopulation in the region of Kings Canyon National Park. We must make two important and independent choices. The first concerns the accuracy with which we want to describe the habitat in question, while the second regards the number and type of variables that we want to use to describe the population dynamics at the local scale.

As outlined by Hanski (1998), there are three qualitatively very different possibilities to describe the habitat (see Fig. 6.3). The simplest alternative is to imagine that the lakes available to the frogs are all equal to each other and arranged in a regular manner (Fig. 6.3a). This description is clearly too simplistic, although it at least preserves the key feature of the territory, that is its fragmentation. The lakes are in fact fragments of habitat that are suitable for the species and immersed in an unsuitable environmental matrix. On the other extreme of the spectrum of possibilities (Fig. 6.3c) one can use a geographic information system (GIS) to detail not only the spatial location and the specific shape of each of the lakes, but also the biotic and abiotic features of the lakes, as well as their riverine connections and a description of the territory that is in between them. Another spatially explicit approach, which is however of intermediate complexity (Fig. 6.3b), is to divide the habitat into a finite number of fragments of regular shape (e.g. circular like in Fig. 6.3b), each of which is characterized by the same size and location as in reality, so that the distances between them can be considered explicitly. Depending on the level of detail at which habitat is modelled, even the movement of organisms between fragments can be

Fig. 6.3 Different ways for describing habitat (top) or demography (bottom) in a simple model for part of the metapopulation of Fig. 6.2 (bottom right corner)

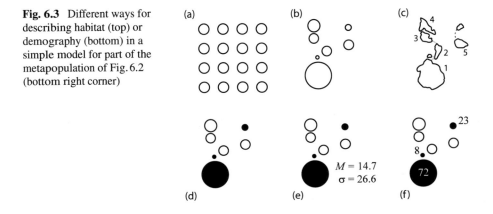

described with different levels of accuracy. Considering the example again, if one imagines that individuals can move with the same probability and effectiveness from any of the lakes to any other (this mechanism is called *propagule rain*), the spatial coordinates of the fragments are not so important and therefore the metapopulation can be described with what is called *spatially implicit* modelling. In this approach the fragments can be accounted for without being located geographically. Whenever the fragments are instead arranged according to precise spatial coordinates, it is more spontaneous to also introduce a realistic description of the organisms dispersal. In these cases, one generally assumes that movement can take place with different probability and effectiveness according to distance and/or direction. The probability distribution of the dispersal distance is called *dispersal kernel*.

Similarly to what was done for the habitat, one can think of three qualitatively different alternatives to model the local demography. The simplest approach considers only the presence and absence of the species *R. muscosa* and produces models that are termed *boolean* (see Fig. 6.3d), because each fragment can be in only one of two states: empty (0) or occupied (1). These models are apt to describe a situation in which the demographic information is very poor, because they exclusively make a distinction between empty fragments and the other. Considering our example again, that lake number 3 is occupied by a single frog or 2–3 adults per coast decameter would not make any difference for a boolean model. Much more informative (Fig. 6.3f) would be an exact count of the number (necessarily discrete) of amphibians inhabiting each lake. This approach, which appears spontaneous, is however much more complex from a mathematical viewpoint because it requires the use of equations similar to those used to deal with demographic stochasticity in Sect. 3.2. An alternative of intermediate complexity may consist in using models that complement the boolean ones by using, for example, information related to the mean and variance of the abundances of frogs in the occupied fragments (Fig. 6.3e). Models in which there exists at least one variable measuring local abundances are called *structured*.

Since the choices pertaining to the description of habitat, dispersal mechanism and demography are relatively independent of each other, many combinations are possible and have therefore been proposed in the literature. The families of available metapopulation models range from those spatially implicit and boolean *à la* Levins (1969), to those that are spatially implicit but structured, such as Markov models (Chesson 1984; Casagrandi and Gatto 1999, 2002) or partial derivatives models (Gyllenberg and Hanski 1992); from models that are spatially explicit and boolean, such as cellular automata (Molofsky 1994; Hiebeler 1997), to the more complex structured and spatially explicit models like e.g. coupled maps (Allen et al. 1993; Hastings 1993; Earn et al. 2000; Yakubu and Castillo-Chavez 2002), or spatially realistic Levins models (Hanski and Ovaskainen 2000) or Interacting Particle Systems (Casagrandi and Gatto 2006). In this chapter, for the sake of simplicity, we will only deal with boolean models, such as the one due to Levins (1969), which still represents a conceptual basis, a sort of "null model", against which to compare predictions deriving from any other more realistic metapopulation model.

6.3   Spatially Implicit Boolean Models

Assume that the metapopulation habitat consists of an infinite number of identical fragments, as shown in Fig. 6.3a. Imagine also that the fragments are arranged in such a way as to make it possible for individuals that disperse from each habitat patch to reach and colonize any other fragment with the same probability and with no time delay due to displacement. This is of course such an abstract idealization of the actual process that it is difficult to find examples of real cases. It is not totally unrealistic, however, especially in cases where the average dispersal distance is relatively large compared to the size of the fragments and the movement takes a time which is much shorter than the average lifetime of the organisms (think for example of the dispersal of light seeds carried by wind). Indicate with $p(t)$ the fraction of patches occupied by the species being studied. This proportion—in the case of a very large number of fragments—can be identified with the probability that any patch is occupied at time t. Quoting from Gotelli (2008), we can say that this single variable is as informative of the state of the metapopulation as the proportion of vacant places in a large parking lot: we content ourselves with knowing how many places are still vacant no matter where they are located. The equation that describes the time change of the variable $p(t)$ is a balance between the fluxes of extinction and occupation of empty fragments:

$$\dot{p}(t) = C'(p(t)) - E'(p(t)) = C(p(t)) \cdot (1 - p(t)) - E(p(t))p(t) \qquad (6.1)$$

where the function $C'(p)$ is the colonization rate (i.e. the fraction of fragments successfully colonized in the unit time) while $E'(t)$ is the rate of extinction (i.e. the proportion of fragments in which the local population becomes extinct in the unit time). The careful reader will recognize a modelling approach which is quite of the same kind as the budget proposed by MacArthur and Wilson (1967) to analyse the variation of the number of species in the context of island biogeography.

The function $C'(p)$ can be described in general terms as the product of the probability of patch colonization per unit time $C(p(t))$ times the fraction of empty fragments $(1 - p(t))$, namely those that can be colonized. Similarly, $E'(p)$ is the product of the probability of extinction per unit time $E(p(t))$ of a local population times the fraction of occupied fragments $p(t)$, namely the only fragments where a local population may become extinct. The forms of dependence on the fraction of occupied fragments p of both the colonization rate (C) and the local extinction rate (E) can differ depending on the type of metapopulation being studied. We now proceed to analysing in more detail two particularly noteworthy cases.

6.3.1   The Mainland-Island Model

Assume that the habitat of the metapopulation is a real archipelago of small islands facing the mainland (as in the diagram of Fig. 6.4a). A landscape of this kind is apt

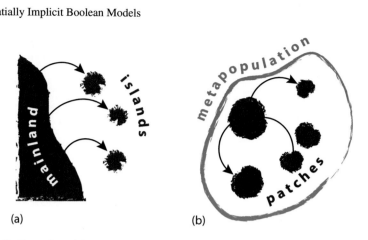

Fig. 6.4 Outlines of a mainland-island metapopulation on the left and a metapopulation *à la* (Levins, 1969) on the right

to characterize not only real archipelagos in the geographical sense, but is useful to describe all those situations in which one of the habitat patches has a much larger size than the others, and a resource availability which is sufficient to make the probability of extinction within the mainland virtually vanishing. Consider, for example, the case of the Edith's checkerspot (*Euphydryas editha bayensis*) in Santa Clara County described in Fig. 6.1, where the large vegetation patch of Morgan Hill effectively acts as a permanent source of individuals for other fragments. As the size of the mainland is incomparably larger than those of the islands, it is reasonable to think that it can host many more individuals of the considered species than each of the small islands. One can imagine, therefore, as a first approximation, that it is possible to neglect, because of that numerical imbalance, any immigration between islands. If the mainland acts as the source of permanent immigration of individuals, the probability rate of an empty patch colonization $C(p)$ can be considered as independent of p, namely a constant equal to c. As for the extinction rate, let us suppose that it is also independent of p namely $E(p) = e$, a constant. This second hypothesis is in some sense equivalent to that of density-independence in birth-death processes. Thus the model (6.1) becomes

$$\dot{p}(t) = c \cdot (1 - p(t)) - ep(t) \tag{6.2}$$

The dynamics of patch occupation as predicted by Eq. 6.2 is easy to study. Equating the right-hand-side of Eq. 6.2 to zero provides only one equilibrium $\bar{p} = \frac{c}{(c+e)}$ which is always feasible, because $0 < \bar{p} < 1$ for any positive value of parameters c and e. This equilibrium is also monotonically increasing with c and is always asymptotically stable, as one can easily prove, for instance graphically. From the conservation viewpoint, we can therefore conclude that the persistence of a mainland-island metapopulation is guaranteed—at least on the theoretical ground—independently of the colonization ability of the considered species and its probability of local extinction. This is obviously due to the mainland providing a perpetual replenishment of organisms to the archipelago of fragments.

6.3.2 The Model by Levins (1969)

Suppose now that there is no mainland to provide a continuous flow of migrating individuals into the patches and that dispersal events are due to migration between fragments (See Fig. 6.4b). This case is obviously much more general than the former and this situation is the one taken into account by Levins (1969). Assume that dispersing individuals from each fragment reach a sort of common reservoir from which they are randomly redistributed to any habitat fragment (the propagule rain paradigm previously described).

The probability of colonization per unit time $C(p)$ can be thought of as growing proportionally to the fraction of occupied fragments p, because the greater p, the greater the amount of dispersing individuals, namely of potential settlers. Note that this assumption is partly arbitrary, because there can be metapopulations in which, although the fraction of occupied fragments is very high, the average number of organisms per fragment is low or vice versa (many empty fragments, but large number of organisms per occupied patch). On the other hand, in boolean models we a priori give up information about the size of the local population and use p as a proxy variable for density, i.e. as an indirect measure of the total abundance of dispersers. Also note that $C(p) = c \cdot p$ vanishes for $p = 0$, as it must indeed be because there is no immigration from outside the metapopulation. The fact that Levins' model does not account for local dynamics does not imply that this cannot be taken into account indirectly. For example, metapopulations with local populations characterized by higher reproductive rates can be described by a larger value of the colonization parameter c.

As for the extinction rate, the simplest assumption is again to set $E(p) = e$ constant. Of course, this is a crude assumption. For instance, one might reasonably wonder whether this parameter might correlate with the dispersal rate (which is actually a component of the colonization parameter c). In fact, dispersal from one patch increases the probability of local extinction, thus implying that in Levins' model the parameters of extinction and colonization might be somehow linked. For the sake of simplicity, we neglect this problem.

We can thus formulate the metapopulation model by Levins as

$$\dot{p}(t) = cp(t) \cdot (1 - p(t)) - ep(t) \tag{6.3}$$

As remarked by Levins (1969) himself in his original paper, the relationship (6.3) is mathematically equivalent to the logistic equation, which describes the dynamics of a single population subject to density dependence. Equation (6.3) can indeed be rewritten as

$$\dot{p} = (c - e)p \cdot \left[1 - \frac{p}{1 - \frac{e}{c}} \right].$$

Therefore, there exist two equilibria of the system. One is mathematically trivial $\bar{p}_0 = 0$ and corresponds to metapopulation extinction. The other one is instead $\bar{p}_+ = 1 - \frac{e}{c}$ and is positive only if $e < c$, i.e. only if the probability of colonization per unit time of an empty patch is greater than the probability of local extinction in the same unit time. Note that the necessary constraint $\bar{p}_+ \leq 1$ is always respected. Indeed equality would apply only if the extinction probability were zero, which never occurs in a real metapopulation. Because of the similarity with the logistic equation, we can conclude that the equilibrium to which the fraction of occupied patches tends whatever the initial condition $0 < p(0) \leq 1$ is given by

$$\bar{p}_+ = \begin{cases} 1 - \dfrac{e}{c} & \text{if } c > e \\ \\ 0 & \text{if } c \leq e \end{cases}$$

A good indicator of the rapidity with which the metapopulation tends towards the equilibrium is the absolute value of the eigenvalue $|c - e|$ of the linearized system, namely the absolute value of the difference between the colonization and the extinction coefficient. Thus the condition for the metapopulation persistence that is obtained from the model (6.3) is very simple: a Levins-like metapopulation persists if and only if the colonization rate exceeds the extinction rate.

6.4 The Effects of Habitat Destruction and Environmental Disasters

As we stated at the beginning of this chapter, it is particularly important not only to predict under what conditions a species hosted by a fragmented landscape can persist, but also how robust its persistence is. In fact, many metapopulations must compensate, via effective dispersal mechanisms, not only for the natural habitat fragmentation, but also for the fragmentation due to man-made land-use change. Also, they must withstand other occasional disturbances originated by a fluctuating environment and/or human activities that threaten their viability (environmental catastrophes).

It is important to comment on the difference between the two different types of disturbance (habitat loss vs. environmental catastrophes). As a prototypical example of environmental catastrophes, think of forest fires, which can occur with different frequencies in different biomes and at intervals more or less regular (Casagrandi and Rinaldi 1999). They may be due to both natural causes (typically heat waves in Mediterranean forests) and human activity. The fire has often the power to wipe out most of the plants of a forest fragment. However, in many cases it does not destroy the habitat of plants. On the one hand, in fact, the dead organic matter deposited to the ground before the fire occurs is more rapidly mineralized. Some chemical elements (such as nitrogen) are lost to the atmosphere as gases, while others (such as calcium and potassium) are instead the main constituents of ashes and can act as fertilizers. If the fire is not too hot and does not give rise to big convective motion, most ashes fall on the burnt site. Therefore, the availability of soluble mineral nutrients may increase and, if they are absorbed by soil, may reduce soil acidity and stimulate primary production. Raising soil temperature may also favour seed germination and root development. In other words, even if the local population of plants in the burnt patch is destroyed, this fragment may still have features that make it suitable to re-colonization. In boolean terms, we can say that the fire event (the environmental catastrophe) switches the state of the fragment from "occupied" to "empty". Similar examples of environmental catastrophe are possible in other ecosystems, of course, for instance the occasional emergence of deadly diseases, the erratic occurrence of very unfavourable microclimatic conditions (very harsh winters or hot summers), or the occurrence of disastrous floods in riverine environments. Grasshoppers of the species *Bryodema tuberculata*, for example, inhabit gravel bars along braided rivers in Central Europe. In these metapopulations the frequency of flood occurrence is fundamental to determine the viability of the species in the river ecosystem (Stelter et al. 1997).

The loss of habitat caused by human activities on natural landscapes have instead very different effects on the persistence of a metapopulation. The conversion of vast portions of pristine tropical forests to agriculture, as well as the construction of buildings and roads, or the channelling of rivers, permanently alters or destroys the habitat fragments involved. By permanently we mean that these actions cause not only the death of individuals that inhabited the fragment before human intervention, but also, so to say, the death of the same fragment, which is thus made unavailable to possible future colonization. Sometimes the destruction is only partial, e.g. due to urbanization, thus leading to habitat erosion and the decrease of the carrying capacity of the involved patches. Although habitat loss directly caused by land-use change is the commonest way of landscape destruction, it is not the only cause. Other important drivers are climate change (as argued in the introduction to Chap. 1) or the deterioration of abiotic conditions (such as the alteration of the availability and abundance of limiting nutrients in the patches). Although it might be important to make a distinction between habitat loss and habitat erosion (see Casagrandi and Gatto 2002b), here we confine the analysis to the effects of the first type of disturbance. Thus we assume that the destruction of a certain fraction of habitat fragments is complete.

Levins' model can easily incorporate both disturbances (habitat loss and environmental catastrophes). Figure 6.5 shows in schematic way how to model the alterations produced by both types of disturbance. If we denote by $h \leq 1$ the fraction of the pristine habitat that is still suitable after the possible destruction made by man, and with $m \geq 0$ the occurrence probability of an environmental disasters per unit time, Levins' model can be modified as follows

Fig. 6.5 Simple scheme that
shows how habitat loss and
environmental catastrophes
can be incorporated into
Levins' model

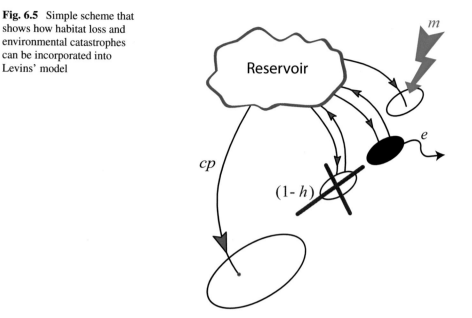

$$\dot{p}(t) = cp(t) \cdot (h - p(t)) - (e + m)p(t) \tag{6.4}$$

The fraction of colonizable landscape, made up by fragments that are empty but
can still be occupied—because they have not been destroyed—is no longer equal
to $1 - p(t)$, like in the original model, but to $h - p(t)$. Dispersers will still try to
colonize a fraction $1 - h$ of the landscape but their dispersal to those locations will
be unsuccessful. As a result of this change, the constraints for the state variable
become $0 \le p(t) \le h$. Note that this implies the assumption that the propagule rain
is purely passive, namely dispersers are not able to actively search for fragments to
colonize. In fact, we assume that individuals move to destroyed patches but fail to
settle there, thus being lost forever. As for the role of environmental catastrophes,
they can be simply modelled by assuming that they raise the rate of extinction of
local populations of a quantity equal to $m \cdot p(t)$.

By requiring that $\dot{p} = 0$ in model (6.4) it is easy to obtain the following equilibrium
conditions

$$\bar{p}_+ = \begin{cases} h - \dfrac{e + m}{c} & \text{if } h > \frac{e+m}{c} \\[2mm] 0 & \text{if } h \le \frac{e+m}{c} \end{cases} \tag{6.5}$$

Similarly to what we previously stated for the undisturbed Levins model, one can
show that the non-trivial equilibrium, when it is feasible, i.e. positive, is also asymptotically stable. It therefore represents the persistence equilibrium of the metapopulation.

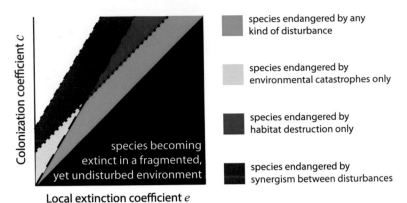

Fig. 6.6 Persistence-extinction boundaries in the Levins model (6.4) which includes anthropic and environmental disturbance. The black region below the solid line corresponds to global extinction of the undisturbed metapopulation as represented by the undisturbed Levins model (6.3). The dotted and the dashed lines, respectively, correspond to the persistence-extinction boundaries in the cases of environmental catastrophes only ($m = 0.1, h = 1$) and habitat loss only ($m = 0, h = 0.7$). The dash-and-dotted line represents the boundary when both disturbances act synergistically

Before analysing the combined effect of environmental catastrophes and habitat destruction, it is reasonable to study the effects independently caused by each of the disturbance parameters introduced in the modified Levins model (6.4). The parameter space in which it is useful to carry on the discussion regarding the effects of the disturbances is certainly the extinction-colonization plane, that is (e, c). Each point in this two-dimensional parameter space can be ideally imagined as representative of a particular metapopulation whose organisms belong to a species living in a specific fragmented territory. As such the metapopulation has its own capacity of colonization (and reproduction, as previously described) and is subject, because of both demographic and dispersal characteristics, to a certain risk of local extinction. The condition of persistence of the undisturbed Levins model (6.3) is represented in the parameter space (e, c) as the bisector of the first quadrant. This line is then what we call the *persistence-extinction boundary*, or the curve in the parameter space that divides regions where persistence is guaranteed from regions where extinction is certain. The points of the black region in Fig. 6.6 thus correspond to metapopulations doomed to become extinct in fragmented landscapes even if they are undisturbed. The region above the bisecting line, instead, represents metapopulations that are guaranteed to persist, if anthropogenic or environmental disturbances are not acting.

We first analyse the effect on persistence of environmental catastrophes only. The condition for equilibrium (6.5) to be positive for $h = 1$ becomes $c > e + m$. Compared to the case of the undisturbed Levins model ($c > e$), the persistence-extinction boundary moves upward in the plane (e, c), as shown in Fig. 6.6. Environmental catastrophes are therefore responsible for the extinction of all the metapopulations whose parameters belong to the region between the black region and the dotted line. The fact that environmental catastrophes imply a simple translation of the persistence-extinction boundary suggests that metapopulations with both high and low extinction rates are likewise influenced by the occurrence of natural or anthropogenic disasters. On the other hand, we have seen in the previous chapter that the critical area of a reserve increases with the diffusion coefficient, in other words, the probability of extinction in a patch of a given area increases with the species dispersal rate. Therefore, in the simple Levins model the dispersal ability of a population is somehow included in the rate of local extinction e; thus, we can conclude that both species that disperse frequently and those that disperse rarely are similarly sensitive to environmental catastrophes.

On the contrary, habitat destruction has a rather selective effect on the likelihood of extinction in different metapopulations. In the absence of environmental disasters ($m = 0$), the positivity and stability condition of equilibrium (6.5) becomes $h > \frac{e}{c}$. As shown by the dashed straight line in Fig. 6.6, the persistence-extinction boundary rotates counter-clockwise in the space (e, c) with respect to the undisturbed situation ($h = 1$). Therefore, metapopulations doomed to extinction uniquely because of habitat loss are those corresponding to the region comprised between the dashed line and the bisector. This region considerably widens for increasing e. We can thus infer that species dispersing more frequently are more affected by this type of disturbance (Casagrandi and Gatto 1999).

Note that, if the two types of disturbance act simultaneously, namely $m > 0$ and $h < 1$, the persistence-extinction boundary becomes the dash-dotted line in Fig. 6.6. Since the set of metapopulations doomed to extinction by the co-occurrence of the two types of disturbance is larger than the union of the two sets obtained earlier, we can conclude that these drivers actually act in a synergistic way, which is thus a phenomenon that contributes to significantly increase the risk of metapopulation extinction.

6.5 A Spatially Explicit Boolean Model

The Levins and other spatially implicit models have the limitation that they cannot be used to analyse explicit spatial patterns. For example, colonization is very often distance-dependent; if this feature is introduced into a model, simulation shows that metapopulation persistence is influenced in a different way by habitat loss (smaller h) occurring randomly or nonrandomly in space.

When it comes to which spatially explicit model to use, there are several possible choices, depending on the kind of available data and the goal of the risk assessment to be conducted. Ideally, one could use a geographical information system to detail the biotic and abiotic characteristics of the landscape continuum. Also, more realistic dispersal descriptions can be used, e.g., via dispersal kernels. An intermediate, yet spatially explicit approach consists in dividing the landscape into a finite number of patches, each of them possibly being characterized by its area, resource availability and location in space (so that the relative distances among the patches can be explicitly included). As for the demography of each population hosted in the patches, the most detailed approach would be the one in which the population dynamics is described by birth-death processes like the ones we introduced in Chap. 3. However, here we will assume that the population dynamics of each landscape fragment can be described in a boolean way (patch is empty or occupied). Perhaps, the best known model of this kind is the one introduced by Hanski and Ovaskainen (2000).

Consider a number n of patches; term $p_i(t)$ the probability that patch i with $i = 1 \ldots n$ is occupied, E_i and C_i the relevant extinction and colonization rate, respectively. Then the rate of change of $p_i(t)$ can be described by a Levins-like model of this kind

$$\dot{p}_i(t) = C_i \cdot (1 - p_i(t)) - E_i \cdot p_i(t). \tag{6.6}$$

Note, however, that $p_i(t)$ is a probability, not a fraction like the one used in the *bona fide* Levins model of the previous sections. The extinction and the colonization rate of population i can be specified in the following way. Let A_i be the area of habitat patch i. The extinction rate is assumed to be proportional to the inverse of patch area, $E_i = e/A_i$ with e being a positive constant, because large patches tend to have large expected population sizes and because extinction risk scales roughly as the inverse of the expected population size (Hanski 1999). Notice that the constant e is measured as area per unit time, like the diffusion coefficient D we introduced in the previous chapter. The colonization rate, instead, is given as a sum of contributions from the existing local populations. Let l_{ji} be the probability that a propagule released by population j reaches patch i; often, it is specified as

$$l_{ji} = L \exp\left(-\alpha d_{ji}\right)$$

where d_{ji} is the distance between the two patches, L is a suitable constant smaller than 1 and $1/\alpha$ is the average dispersal distance. This is an example of an exponential dispersal kernel (but other choices are possible, e.g. a Gaussian kernel, Clark et al. (1999)). If l_{ji} only depends on distance, then $l_{ji} = l_{ij}$, i.e. the dispersal probability matrix is symmetric. Then the colonization rate is given by

$$C_i = c \sum_{j \neq i} l_{ji} A_j p_j(t)$$

because immigration to patch i is expected to increase with the number of individuals in neighbouring populations, as reflected by the respective patch areas, and with their decreasing distances to the focal patch and increasing probability of occupancy. The constant c is positive and is a probability of colonization per unit area per unit time.

Thus, the dynamics of the metapopulation is given by

$$\dot{p}_i(t) = (c/A_i) \left[\sum_{j \neq i} m_{ji} p_j(t) \cdot (1 - p_i(t)) - (e/c) p_i(t) \right]. \tag{6.7}$$

where $m_{ji} = l_{ji} A_i A_j = m_{ij}$ are the elements of a matrix M, which we will term the migration matrix from now on. Model (6.7) can admit several possible equilibria, but the persistence of the metapopulation can be anyway established by examining the stability of the extinction equilibrium, namely $p_i = 0$ $\forall i$. In order to do that, we linearize the model around the zero equilibrium. This amounts to discarding the quadratic terms in the right-hand-side of Eq. 6.7, namely

$$\dot{p}_i(t) = (c/A_i) \left[\sum_{j \neq i} m_{ji} p_j(t) - (e/c) p_i(t) \right]. \tag{6.8}$$

Equation 6.8 can be rewritten with matrix notation as

$$\dot{p} = cA^{-1}\left(M - \frac{e}{c}I\right)p$$

where

$$p = \begin{bmatrix} p_1 \\ p_2 \\ \cdot \\ \cdot \\ p_n \end{bmatrix} \quad A = \begin{bmatrix} A_1 & 0 & . & . & 0 \\ 0 & A_2 & 0 & . & 0 \\ . & . & . & . & . \\ . & . & . & . & . \\ 0 & 0 & 0 & . & A_n \end{bmatrix} \quad M = \begin{bmatrix} 0 & m_{12} & m_{13} & . & m_{1n} \\ m_{12} & 0 & m_{23} & . & m_{2n} \\ . & . & . & . & . \\ . & . & . & . & . \\ m_{1n} & m_{2n} & m_{3n} & . & 0 \end{bmatrix},$$

M is the migration matrix, and I is the identity matrix. Since M is a symmetric matrix, its eigenvalues are real. The largest eigenvalue is termed the dominant eigenvalue. Hanski and Ovaskainen (2000) have shown that the extinction equilibrium is unstable, so that the metapopulation can persist, if and only if the dominant eigenvalue λ_M of M is larger than e/c:

$$\lambda_M > \frac{e}{c}. \tag{6.9}$$

λ_M is measured as a squared area and has been aptly defined *metapopulation capacity* by Hanski and Ovaskainen (2000). Persistence is thus guaranteed (Eq. 6.9) if the metapopulation capacity exceeds the threshold e/c. Note that it is possible to recover the result of the simple Levins model. It is sufficient to assume that all patches are equal, have the same unitary area, and there is a rain of propagules (i.e. $l_{ji} = 1/(n-1)$); in that case, as we already know, the condition for persistence is $1 > e/c$. In other words the metapopulation capacity for that model is 1.

The big advantage of a spatially explicit model is that one can specify the location and the area of the habitat patches and simulate the fate of the corresponding metapopulation. It is easy to introduce realistic patterns of habitat loss and environmental catastrophes by manipulating the model parameters. For instance, habitat erosion, that is a proportional reduction of all the habitat patches, can be simulated by assuming that the areas of all the patches are decreased by a certain proportion. Consequently, the migration matrix M changes, its dominant eigenvalue can be recal-

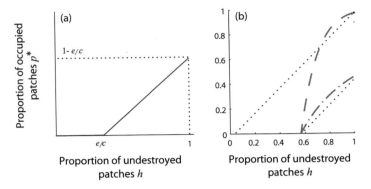

Fig. 6.7 Fraction of occupied patches at equilibrium in a hypothetical metapopulation as a function of the preserved habitat fraction h calculated by means of different models. **a** Modified Levins model (6.4); **b** Spatially implicit structured model (Casagrandi and Gatto, 2002b) in the case of high (orange, dashed curve) and low (blue, dash-and-dotted curve) dispersal frequency. Dotted curves correspond to the results that would be obtained by utilizing the simple model by Levins (1969)

culated and compared to the threshold e/c to see whether persistence is possible. Of course, simulation of the differential Eq. 6.7 allows one to determine also the probability that each specific habitat patch is occupied in both the short and the long term. On the other hand, strategies that aim at the preservation of a species can also be simulated. The establishment of an ecological corridor between patch i and patch j, for instance, can be mimicked by an increase of the relevant dispersal coefficient l_{ji}.

The conclusions obtained in this chapter are subject to a number of limitations related to the excessive simplicity of boolean models from which results were derived. The reader should keep this consideration in mind as a guide to planning and managing the conservation of metapopulations. In particular, more sophisticated models (for a better description of either habitat or local demography) should be used if one seeks for solutions to be practically implemented. Sometimes, in fact, boolean models significantly underestimate the extinction risk of metapopulations. An important example in this respect is related to the estimate of the proportion of habitat that must be preserved so that a metapopulation may persist even when part of the habitat is destroyed. The simplicity of the laws derived from the boolean model (6.4) is indeed very tempting and the risk of using it in decision-making contexts is considerable. Figure 6.7a shows the fraction of occupied patches in a hypothetical metapopulation as a function of the preserved habitat fraction h calculated by means of the modified Levins model. As one can note, this relationship is linear and vanishes for $\hat{h} = \frac{e}{c}$. Therefore, following the directions of the boolean model, a decision-maker who has to grant permissions for exploiting the land for other purposes (for example, converting forested fragments into agricultural use, or building resorts in an island archipelago) could be led to believe that a proportion of habitat can be exploited to the limit of $1 - \hat{h} = 1 - \frac{e}{c}$, i.e. exactly the fraction of habitat patches occupied by the species before land-use change. This "rule of thumb" for environmental conservation, although quite popular because of its simplicity, is very gross and should not be used in practice without careful consideration. Figure 6.7b displays the proportion of occupied fragments at equilibrium as a function of h in a more sophisticated metapopulation model, which includes local population dynamics and demographic stochasticity, in two different cases. While the dot-and-dashed curve, which refers to a species with low frequency dispersal, rather faithfully reproduces the result that is obtained via the Levins model, the dashed curve (referring to a species with a high frequency dispersal) displays a very different result. In fact, destruction of 40% of the habitat is sufficient to condemn such a metapopulation to extinction even if more than 90% of the fragments were occupied before land-use conversion. Therefore, it is good to remind the reader that, before translating the simple management policies suggested in this chapter into concrete implementation, it is necessary to evaluate the quantitative effects with more appropriate tools than simple boolean models. Available PVA software like RAMAS and VORTEX can help the ecologists to carry out the job.

References

Allen JC, Schaffer WM, Rosko D (1993) Chaos reduces species extinction by amplifying local population noise. *Nature* 364:229–232

Casagrandi R, Gatto M (1999) A mesoscale approach to extinction risk in fragmented habitats. *Nature* 400:560–562

Casagrandi R, Gatto M (2002) A persistence criterion for metapopulations. *Theoretical Population Biology* 61:115–125

Casagrandi R, Gatto M (2002) Habitat destruction, environmental catastrophes, and metapopulation extinction. *Theoretical Population Biology* 61:127–140

Casagrandi R, Gatto M (2006) The intermediate dispersal principle in spatially explicit metapopulations. *Journal of Theoretical Biology* 239:22–32

Casagrandi R, Rinaldi S (1999) A minimal model for forest fire regimes. *American Naturalist* 153(5):527–539

Chesson PL (1984) Persistence of a Markovian population in a patchy environment. *Zeitschrift für Wahrscheinlichkeitstheorie* 66:97–107

Clark JS, Silman M, Kern R, Macklin E, HilleRisLambers J (1999) Seed dispersal near and far: Patterns across temperate and tropical forests. *Ecology* 80:1475–1494

Curran LM, Trigg SN, McDonald AK, Astiani D, Hardiono YM, Siregar P, Caniago I, Kasischke E (2004) Lowland forest loss in protected areas of Indonesian Borneo. *Science* 303:1000–1003

Earn DJD, Levin SA, Rohani P (2000) Coherence and conservation. *Science* 290:1360–1364

Ehrlich PR (1965) The population biology of the butterfly *Euphydryas editha*. ii. The structure of the Jasper Ridge colony. *Evolution* 19:327–336

Fearnside PM (2001) Soybean cultivation as a threat to the environment in Brazil. *Environmental Conservation* 28:23–38

Gotelli NJ (2008) *A Primer of Ecology*, 4th edn. Sinauer, New York

Gyllenberg M, Hanski I (1992) Single-species metapopulation dynamics: A structured model. *Theoretical Population Biology* 42:35–61

Hanski I (1998) Metapopulation dynamics. *Nature* 396:41–50

Hanski I (1999) *Metapopulation Ecology*. Oxford University Press, Oxford

Hanski I, Ovaskainen O (2000) The metapopulation capacity of a fragmented landscape. *Nature* 404:755–758

Harrison S, Murphy DD, Ehrlich PR (1988) Distribution of the bay checkerspot butterfly, *Euphydryas editha bayensis*: Evidence for a metapopulation model. *American Naturalist* 132:360–382

Hastings A (1993) Complex interactions between dispersal and dynamics: Lessons from coupled logistic equations. *Ecology* 74:1362–1372

Hiebeler D (1997) Stochastic spatial models: from simulations to mean field and local structure approximations. *Journal of Theoretical Biology* 187:307–319

Knapp RA, Boiano DM, Vredenburg VT (2007) Removal of nonnative fish results in population expansion of a declining amphibian (mountain yellow-legged frog, *Rana muscosa*). *Biological Conservation* 135:11–20

Levins R (1969) Some demographic and genetic consequences of environmental heterogeneity for biological control. *Bulletin of the Entomogical Society of America* 15:237–240

Levins R (1970) Extinction in some mathematical questions in biology. *Lecture Notes on Mathematics in The Life Sciences*. The American Mathematical Society, Providence, RI, pp 75–107

MacArthur RH, Wilson EO (1967) *The Theory of Island Biogeography*. Princeton University Press, Princeton, NJ, USA

Molofsky J (1994) Population dynamics and pattern formation in theoretical populations. *Ecology* 75:30–39

Parmesan C (1996) Climate and species' range. *Nature* 382:765–766

Stelter C, Reich M, Grimm V, Wissel C (1997) Modelling persistence in dynamic landscapes: lessons from a metapopulation of the grasshopper *Bryodema tuberculata*. *Journal of Animal Ecology* 66:508–518

Vredenburg VT (2004) Reversing introduced species effects: Experimental removal of introduced fish leads to rapid recovery of a declining frog. *Proceedings of the National Academy of Sciences of the United States of America* 101:7646–7650

Yakubu AA, Castillo-Chavez C (2002) Interplay between local dynamics and dispersal in discrete-time metapopulation models. *Journal of Theoretical Biology* 218:273–288

Chapter 7
Problems on Spatial Ecology

Problem SE1

The ragweed (genus *Ambrosia*) is an alien species that has been unfortunately introduced to both Europe and Asia from North America. Ragweed pollen is notorious for causing allergic reactions in humans, specifically allergic rhinitis. Therefore, many countries have decided to employ biological control against ragweed, specifically by a leaf-eating beetle, *Ophraella communa* Le Sage, which is quite effective.

Yamamura et al. (2007) studied the mobility of *Ophraella communa* in Japan. On August 8, 2000, they released a number of beetles which they have recaptured after 20 h by means of traps located at various distances. Here below you can find a table with the results in terms of fractions of caught beetles (for instance, 7.6% is the percentage of beetles recaptured between 20 and 30 m).

Distance (m)	0–10	10–20	20–30	30–40	40–50	50–60	60–70	70–80	80–90
Fraction (%)	50	14.5	7.6	10	5.2	4.8	3.2	2.4	2

Assume that beetles dispersed into the environment according to an isotropic two-dimensional model of diffusion. Estimate the diffusion coefficient D.

Problem SE2

Since the beginning of 1900 the Eurasian collared dove (*Streptopelia decaocto*) has expanded its spatial range from Turkey to the whole Europe so that in 1970 Scandinavia, British Isles and Russia have been reached by the bird. The exponential increase of the total number $P(t)$ of doves in a sequence of successive years is displayed in the Figure. Estimate the instantaneous rate of Malthusian increase r of *S. decaocto*.

© The Author(s), under exclusive license to Springer Nature Switzerland AG 2022
M. Gatto and R. Casagrandi, *Ecosystem Conservation and Management*,
https://doi.org/10.1007/978-3-031-09480-4_7

Problem SE1 The defoliator *Ophraella communa*

Problem SE2 Data on the diffusion of collared doves in Europe

It has been further estimated that the range expansion speed for this dove has been 43.7 km year^{-1}. From this information estimate the diffusion coefficient D of this alien species.

Problem SE3

The giant hogweed (*Heracleum mantegazzianum*) has spread rapidly in a number of European countries after introduction as an ornamental from its native area in Russia and Georgia. It is an undesirable invader on account of its large size, prolific seed production and vigorous growth leading to gross changes in vegetation, obstruction of

Problem SE3 The giant
hogweed

*Heracleum
mantegazzianum*

access to river banks, soil erosion, and serious dermatological effects on skin contact.
Pyšek (1991) reports the following data on the invasion of the Czech Republic:

Year	1950	1955	1960	1970	1975	1980	1985
Area (km²)	71	85	136	216	407	512	1019

Nehrbass et al. (2006) have calculated a finite annual growth rate equal to 1.12.
Assume that the areas are approximately circular and estimate the diffusion coefficient of the giant hogweed in Czechia.

Problem SE4

The Himalayan balsam (*Impatiens glandulifera*) was introduced into Europe from the Himalayas as a garden ornamental and nectar-producing plant in the first half of the 19th century. Since then it has invaded nearly all European countries.

Pyšek (1991) reconstructed the alien plant invasion of the former Czechoslovakia geographical area. The graph in the Figure reports the distance travelled by the plant while expanding along river habitats. Also, Pyšek and Prach (1995) report that the demographic growth rate of the invasive species in riparian habitats is $0.073\,year^{-1}$. From this information calculate the coefficient of diffusion D of *I. glandulifera*.

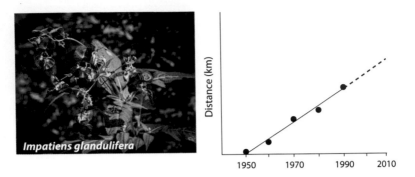

Problem SE4 The distance travelled by the Himalayan balsam (*Impatiens glandulifera*) in Czechoslovakia as a function of years. Modified after Pyšek and Prach (1995)

Problem SE5

The American mink (*Neovison vison*) is a small carnivore of the mustelid family. It is an alien species to Europe where it was imported for fur farming. In Scotland fur farms were first established in 1938 and mink were first recorded breeding in the wild in 1962. Then the carnivore expanded its range.

Fraser et al. (2015) report the following information on mink expansion. In west Scotland, linear expansion rate was in the average 13.8 km year^{-1}, while on the coast in the far north-east of Scotland, linear expansion rate was in the average 2.6 km year^{-1}. The demography of mink is as follows: 0.6 female offspring per adult female per year, juvenile survival up to adulthood approximately equal to 0.9, adult survival between years approximately equal to 0.75. From these data estimate the diffusion coefficients of mink in the west and the north-east of Scotland.

Problem SE5 The American mink

Problem SE6

A new reserve is being set up to protect the species *Perdix polyvarians*, a fantasy bird threatened with extinction. Scientists estimated that the yearly finite rate of Malthusian increase for *P. polyvarians* is 1.05, while the diffusion coefficient is 9 km^2/year. Calculate the critical reserve size for the circular and the square shape.

Problem SE7

The sea otter (*Enhydra lutris*) was on the verge of extinction in California, because of overharvesting, but fortunately in 1911 an international treaty was signed in order to protect the species (Lubina and Levin 1988). Its numbers have been exponentially increasing, as shown in the table below

Year	1938	1947	1950	1955	1957	1959	1963	1966	1969	1972
No. of otters	310	530	660	800	880	1050	1190	1260	1390	1530

The sea otter has not only increased its abundance but has also expanded its range along the Californian coastline with a speed of about 2.2 km year^{-1}. From this information estimate the diffusion coefficient D of the otter, assuming that the habitat (coastline) is one-dimensional.

Suppose then that you want to introduce a new population of *E. lutris* into the country of Whatsoever, where only a piece of shoreline, 80 km long, is the right habitat for hosting *E. lutris*. Outside that piece of shoreline the sea otter cannot survive at all. Assuming that the new population shares the same demographic and dispersal parameters with the Californian population, assess whether the new population can be successful or not.

Problem SE7 The sea otter

Enhydra lutris

Dreissena polymorpha

Problem SE8

The zebra mussel (*Dreissena polymorpha*) is a freshwater invasive species which in
the past 25 years has been responsible for enormous damages, both environmental and
economic, specially in the United States (and marginally in Italy too). The average
lifetime of *D. polymorpha* is about 4 years, and each female produces in the average
2×10^5 eggs per year, with sex ratio 1:1. The fraction of eggs surviving up to adulthood
is very low, approximately equal to 0.005%. Adult individuals are basically sessile,
while larvae can be transported by the river stream for a few days, after which they
settle in the river or lake bed and become mature adults that can reproduce.

Recently, the species has been accidentally introduced into the Yellowish River,
downstream of the dam of Deep Lake. The downstream expansion speed of the
species has been estimated to be 640 km/year. You know that the river current implies
a drift of the larvae of about 600 km/year. Estimate the diffusion coefficient of *D.
polymorpha*, assuming that the spatiotemporal dynamics of the population is well
described by a model including advection, diffusion and Malthusian growth.

Problem SE9

The Himalyan thar (or tahr, *Hemitragus jemlahicus*), a large ungulate native to the
Himalayas, was introduced into New Zealand in 1904. It then spread through the
Southern New Zealand Alps. Caughley (1985) has estimated the areas occupied by
thar in subsequent years:

Year	1936	1946	1956	1966	1976	1984
Area (km^2)	129	542	1237	3998	6138	4937

Caughley (1970) provides the following information on thar demography:

Problem SE9   The Himalyan
thar

Hemitragus jemlahicus

- The average numbers of kids produced by one adult female per year is 0.9;
- Sex ratio at birth is 1:1;
- The fraction of kids surviving to adulthood is 0.6;
- The average lifetime of thar is 8 years.

Assume that the areas occupied by the ungulate are approximately circular. Then calculate:

(a) the average expansion speed of thar in New Zealand;
(b) its Malthusian instantaneous rate of demographic increase;
(c) its diffusion coefficient D.

Problem SE10

The European tree frog (*Hyla arborea*) although it is arboreal does anyway need water bodies for reproduction. Carlson and Edenhamn (2000) studied a metapopulation of this frog in Sweden (378 ponds over an area of about 1200 km^2) and estimated in subsequent years the fraction of occupied ponds and the rate of colonization, namely the fraction of empty ponds being colonized in the course of each year, as shown in the Figure.

Assume that the values recorded in 1992 can be considered as values at equilibrium. Use the Levins model in order to estimate the extinction parameter e and the colonization parameter c. If 40% of the ponds were dried up, what effect would this disturbance have on the frog metapopulation?

Problem SE10   Data of a Swedish metapopulation of *Hyla arborea* between 1989–1992. **a** Fraction of occupied ponds in subsequent years. **b** Fraction of ponds that were empty in the previous year and turn out to be occupied in the current year

Problem SE11

You are required to study the metapopulation of the fantasy butterfly species *Euphydryas imaginifica* in the Tuahutu archipelago, which consists of very many small islands. The local extinction probability per unit time in each island is 0.05 year^{-1}. Extinction is counterbalanced by immigration, which can take place because of migration from the mainland or from surrounding islands where *E. imaginifica* is still present. The colonization rate of an empty island from mainland is 0.01 yr^{-1}, while the colonization rate from other islands follows the Levins model and is 0.045 yr^{-1}. Write down the metapopulation model that governs the dynamics of the fraction of occupied islands. Calculate the fraction of occupied islands at equilibrium.

Suppose then that tourism development plans are devised that aim at urbanizing both the Tuahutu archipelago and the nearby coastline on the mainland. According to plan A the archipelago is heavily urbanized, thus leading to the destruction of one fourth of the insular habitat of *E. imaginifica*; according to plan B it is the coast of the mainland that is heavily urbanized, thus leading the whole butterfly mainland population to the complete destruction. Evaluate the impacts of both plan A and plan B on the fate of the butterfly metapopulation.

Problem SE12

Colonization rate c in metapopulations is generally a decreasing function of the average distance between patches that constitute the metapopulation habitat. Therefore, a further impact of habitat destruction is the increase of the average distance between patches, which implies the decrease of the colonization rate. Include this further effect into the Levins model with habitat destruction that has been illustrated previously. Assume that the metapopulation you are studying is characterized by an extinction rate $e = 0.2$ year^{-1}, that $1/4$ of the original habitat is being destroyed and that this destruction implies an increase of the average distance d from 1 km to 1.3 km. You have estimated that c [year^{-1}] $= 0.4/d^2$. Explain whether the metapopulation can persist in the long term after the habitat has been destroyed. If it can persist, estimate the fraction of occupied patches at equilibrium.

Problem SE13

van der Merwe et al. (2016) have studied a metapopulation of rice rats (*Oryzomys palustris*) in a Mississippi floodplain site located in Illinois.

Data on the dynamics of the frequency of occupied patches towards equilibrium in that floodplain are reported in the Figure. The authors have also been able to find a relationship between the extinction rate coefficient e and the 3-month average rainfall (see attached graph). Instead, the colonization rate coefficient c seems to be rather independent of rainfall. In the Illinois location the rainfall is about 200 mm.

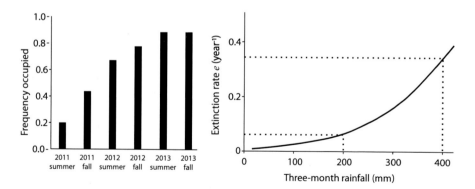

Problem SE13 The dynamics of the fraction of patches occupied by the rice rate in the Illinois floodplain (left); relationship between extinction rate and 3-month average rainfall (right). Modified after van der Merwe et al. (2016)

(a) estimate c from the data in the Figure;
(b) suppose that in another floodplain location the rainfall is 400 mm and that the colonization rate coefficient c is still the same; estimate whether this metapopulation can persist and, if it persists, find the occupation frequency at steady state.

Problem SE14

The Amazon forest has been both destroyed and fragmented as exemplificatively shown in the Figure. This has serious consequences on the viability of many bird species. Ferraz et al. (2007) have studied the dynamics of these metapopulations. In particular for the white-chinned woodcreeper (*Dendrocincla merula*) and the black-throated antshrike (*Frederickena viridis*) they provide the following data that refer to distinct metapopulations each being characterized by a different patch size.

Species name	Patch size (hectares)	Extinction rate e (year^{-1})	Patch occupancy at equilibrium (%)
	1	0.89	5
Dendrocincla merula	10	0.85	7
	100	0.45	10
	1	0.72	20
Frederickena viridis	10	0.65	23
	100	0.05	82

For each of the six metapopulations (each metapopulation is isolated from the other metapopulations and from the continuous Amazon forest) calculate the colonization rate coefficient c via the Levins model.

Ferraz et al. (2007) have also studied the same kinds of metapopulations for *D. merula* when they are in close contact with the continuous Amazon forest. The forest acts as a reservoir of organisms and provides a continuous flow f of birds

Problem SE14 Aerial view of Amazon forest fragmentation in Brazil

to the patches of the metapopulations which is thus added to the colonization flow between patches. Of course, Ferraz et al. (2007) have recorded higher occupancy at equilibrium with respect to the previous situation. The results are reported here below.

Patch size (hectares)	Patch occupancy at equilibrium (%)
1	83
10	85
100	95

Calculate the flow f for each of the three metapopulations.

Problem SE15

In the Levins metapopulation model with habitat destruction, the landscape is divided into three segments: if h is the fraction of landscape still suitable after destruction, then 1-h is the fraction of landscape which is unsuitable, $h - p$ the fraction which is suitable and unoccupied and p the fraction of landscape which is suitable and occupied by individuals of the considered species. Dispersal is assumed to be totally random so that of the flow cp of dispersers only the fraction $cp(h - p)$ will actually colonize empty patches and contribute to the increase of the fraction of occupied patches. In some cases, however, the flow $cp(1 - h)$ of dispersers arriving at the unsuitable habitat is not completely lost. For example, the seeds released by a plant and landing on unsuitable soil can be intercepted by animals that will carry them where the habitat is suitable. In this way a fraction $\alpha cp(1 - h)$ with $0 < \alpha < 1$ will actually be transported to empty patches and contribute to the increase of p.

(A) Add the term $\alpha cp(1 - h)$ into Levins equation with habitat destruction and obtain the relationship that links the fraction of occupied landscape p at equilibrium to the remaining habitat h.

(B) The aquatic plant *Eichhornia paniculata* (Brazilian water hyacinth) studied by Purves and Dushoff (2005) is an example of this kind. Waterfowl and cattle can transport seeds of the plant to the suitable pools where the plant can survive. The Figure shows how the fraction of occupied landscape varies as a function of the fraction of suitable landscape h. Assume that p is at equilibrium and that the extinction of a local population of hyacinths in a pool occurs on average once a year.

Estimate the values of α and c.

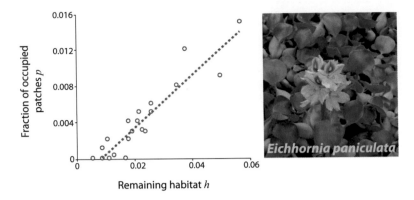

Problem SE15 The fraction of landscape occupied by the Brazilian water hyacinth as a function of the fraction of suitable landscape. Modified after Purves and Dushoff (2005)

Problem SE16

Consider a simple metapopulation consisting of two patches linked by migration. The area of patch 1 is $1\,\text{km}^2$, that of patch 2 is $4\,\text{km}^2$. The probability that a propagule released by patch i reaches patch j is $l_{ij} = 0.37 \exp\left(-\alpha d_{ij}\right)$ where $\alpha = 0.3\,\text{km}^{-1}$ and $d_{12} = d_{21} = 5$ km. The colonization rate is $c = 0.15\,\text{km}^{-2}\,\text{year}^{-1}$, while the extinction rate is $e = 0.1\,\text{km}^2\,\text{year}^{-1}$.

Write down the two Levins-like equations describing the dynamics of p_1 and p_2, namely the probabilities that patch 1 and patch 2 are occupied. Analyze the equations via the isocline method and find out the fate of the metapopulation. If it can persist, calculate p_1 and p_2 at equilibrium. Calculate the metapopulation capacity and verify the condition for metapopulation persistence.

Problem SE17

Consider the 3-patch metapopulation shown in the Figure, where patch 2 and patch 3 are linked to patch 1, but patch 2 and 3 are not connected by migration.

More precisely, the probability l_{ij} that a propagule released by patch i reaches patch j is detailed as follows: $l_{23} = l_{32} = 0$, $l_{12} = l_{21} = 0.027 \exp\left(-\alpha d_{12}\right)$, $l_{13} = l_{31} = 0.09 \exp\left(-\alpha d_{13}\right)$ with $\alpha = 0.4\,\text{km}^{-1}$. The areas of the patches are $A_1 = 10\,\text{km}^2$, $A_2 = 7\,\text{km}^2$, $A_3 = 4\,\text{km}^2$. The colonization rate is $c = 0.1\,\text{km}^{-2}\,\text{year}^{-1}$, while the extinction rate is $e = 0.08\,\text{km}^2\,\text{year}^{-1}$. Calculate the metapopulation capacity and check whether the condition for metapopulation persistence is verified. Assume that ecological corridors are built to improve the migration between 1 and 2 and between 1 and 3. This implies a decrease of α from $0.4\,\text{km}^{-1}$ to $0.25\,\text{km}^{-1}$. Check the condition for persistence again.

Problem SE17 Scheme of
the 3-patch metapopulation

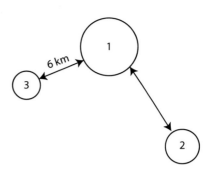

References

Caughley G (1970) Eruption of ungulate populations, with emphasis on Himalayan thar in New Zealand. *Ecology* 51:53–72

Caughley G (1985) A reappraisal of the distribution and dispersal of female Himalayan thar in New Zealand. *New Zealand Journal of Ecology* 8:5–10

Ferraz G, Nichols JD, Hines JE, Stouffer PC, Bierregaard RO, Lovejoy TE (2007) A large-scale deforestation experiment: Effects of patch area and isolation on Amazon birds. *Science* 315:238–241

Fraser EJ, Lambin X, Travis JMJ, Harrington LA, Palmer SCF, Bocedi G, Macdonald DW (2015) Range expansion of an invasive species through a heterogeneous landscape—The case of American mink in Scotland. *Diversity and Distributions* 21:888–900

Lubina J, Levin S (1988) The spread of a reinvading species: Range expansion in the California sea otter. *American Naturalist* 131:526–543

Nehrbass N, Winkler E, Pergl J, Perglová I, Pyšek P (2006) Empirical and virtual investigation of the population dynamics of an alien plant under the constraints of local carrying capacity: *Heracleum mantegazzianum* in the Czech republic. *Perspectives in Plant Ecology, Evolution and Systematics* 7:253–262

Purves DW, Dushoff J (2005) Directed seed dispersal and metapopulation response to habitat loss and disturbance: Application to *Eichhornia paniculata*. *Journal of Ecology* 93:658–669

Pyšek P (1991) *Heracleum mantegazzianum* in the Czech republic: Dynamics of spreading from the historical perspective. *Folia Geobotanica et Phytotaxonomica* 26:439–454

Pyšek P, Prach K (1995) Invasion dynamics of *Impatiens glandulifera*—a century of spreading reconstructed. *Biological Conservation* 74:41–48

van der Merwe J, Hellgren EC, Schauber EM (2016) Variation in metapopulation dynamics of a wetland mammal: The effect of hydrology. *Ecosphere* 7:e01275

Yamamura K, Moriya S, Tanaka K, Shimizu T (2007) Estimation of the potential speed of range expansion of an introduced species: Characteristics and applicability of the gamma model. *Population Ecology* 49:51–62

Part III
Sustainabilty of Biomass Harvesting and Its Management

Chapter 8
The Management of Natural Populations Harvesting

8.1 Renewable Resources

One of the most important problems of applied ecology is certainly the rational utilization of natural populations on the part of man. Hunting, fishing, forestry have always been a source of food, energy, fibre, timber, and any sort of substances useful to humans. In the past century these activities have reached such a dimension as to raise a number of concerns: extinction of species vulnerable to overexploitation, scarce profitability of mismanaged natural resources, difficulty in coordinating activities that take place on a large or very large scale (sometimes involving many countries). The purpose of this chapter is to illustrate the main concepts that are the basis of proper planning and management of the exploitation of those resources that, for their inherent ability to reproduce, are usually called *renewable resources*. The main features that make these resources very peculiar, in particular that make it difficult to introduce more rationality in their exploitation, are basically two.

The first feature is inherent to their being *renewable* and lies in the fact that an increasing effort in exploiting them by humans implies in the long run a smaller production. In fact, if a natural population is greatly reduced in biomass, its rate of increase, namely its production of new biomass in the unit time, is also greatly reduced thus implying small harvests and low profitability of the economic activity in the long term. This clearly distinguishes the exploitation of renewable resources from other economic enterprises in which larger investment and labour employment—i.e. greater production effort—always implies larger production. A particularly relevant example in this sense can be provided by salmon fishing in Alaska during the late nineteenth and the first half of the twentieth century.

Commercial exploitation of salmon in Alaska began around 1880. Two of the five salmon species found in the waters of Pacific Ocean are the most important for commercial fishing: the *sockeye* salmon (*Oncorhynchus nerka*) and the *pink* salmon (*Oncorhynchus gorbuscha*). The pacific salmon is an anadromous species, namely it mainly lives in salt water but breeds in fresh water. Each population is linked to a specific river, because salmon have the unbelievable capacity of returning as adults

© The Author(s), under exclusive license to Springer Nature Switzerland AG 2022
M. Gatto and R. Casagrandi, *Ecosystem Conservation and Management*,
https://doi.org/10.1007/978-3-031-09480-4_8

to the same river where they were born. It is in fact in the river estuaries that they are mainly harvested when they return to reproduce. Most of the catch is canned, thus one can follow the development of this fishery by the number of cases of canned salmon from Alaska. The top panel of Fig. 8.1 displays a slow growth from 1878 to a peak of production of 8 million and a half cases in 1936. Since then, the quantity of harvested salmon has been steadily decreasing until 1960. Of course, there has been no diminishing demand for salmon; in fact the fishing flotilla has become gradually larger and larger (see the green curve in the lower panel of Fig. 8.1) and the number of operating boats has increased by an order of magnitude between 1907 and 1955.

Therefore, an increasing number of fishermen has been catching a decreasing number of salmon. The size of the occurred collapse is evidenced by Fig. 8.1b which reports (in red) the trend over the years in the number of salmon caught by each boat. While in 1908 each boat's harvest averaged 15,000 salmon, in 1954 each boat was fishing in the average only 1,500. Since the catch of each boat is more or less proportional to the number of salmon present in the estuary, it follows that Alaska salmon populations have been reduced by an order of magnitude in less than fifty years.

This disaster is not due to worsening conditions of the environment in which the salmon lives. In fact in neighbouring British Columbia (Canada), there was no

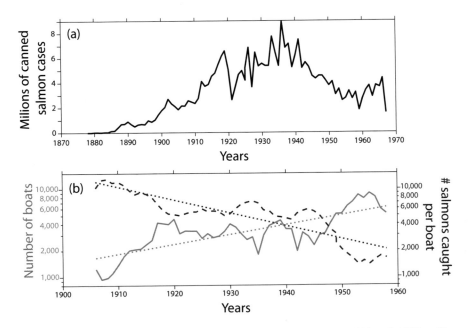

Fig. 8.1 Data pertaining to salmon fishing in Alaska (USA) between 1878 and 1967. **a** Time evolution of the total number of canned salmon cases produced (Byerly et al. 1999). **b** Number of salmon fishing boats (green) and average number of salmon caught by one boat (red) between 1906 and 1959 (data after Cooley 1963); scale is logarithmic and dotted lines display regression versus time

decline in the salmon catch. And the existence of different salmon populations in the region has indeed allowed the restoration of fishing in the area in recent years (Hilborn et al. 2003). Without doubt what is to blame is having allowed the size of the flotilla operating in this fishery to increase dramatically, thus subjecting the Alaska salmon populations to overexploitation. The decimated salmon populations of the Alaska rivers cannot be so productive as they could have been if the exploitation effort had been kept to a lower level.

The second feature that makes renewable resources so peculiar is not always present, but is certainly very frequent, and is their being *open access*. This term means that they are not owned by an individual institution or company or, in the limit, a single person, rather they can be harvested by anyone willing to do it, in other words they belong to the economic operators that first appropriate them. There are important exceptions, though, such as fish farms owned by fishing cooperatives or individuals, or privately owned forests, or hunting reserves managed by hunters' associations. However, very frequent is the case of resources that are located in state-owned territory or waters and can be exploited by anyone, possibly with the necessary rules and due licenses. Not to mention some enormous fish stocks that are not located in the territorial waters of any nation or are located in the territorial waters of several countries: in the first case open access is the only possible rule, in the second it is often the only rule that makes it practicable to find a diplomatic agreement among the interested countries.

The essence of the difficulty inherent to rationalizing the exploitation of natural populations is substantially represented by the two characteristics of being renewable and open access. A hypothetical, though realistic, example due to Colin (Clark 1981), effectively illustrates it. Clark calls the hypothetical example the fishermen's *dilemma* by analogy with the prisoners' dilemma well known in game theory (Luce and Raiffa 1967), but it might be equally termed the hunters' or lumberjacks' dilemma.

Suppose you consider a renewable resource with two competing exploiters. By way of example, let us refer to two fishermen operating on opposite sides of the same stretch of river. We denote for simplicity the two fishermen with the letters A and B. Suppose that A and B have two possible strategies only:

- conserve by moderately exploiting the resource (for example by fishing one day a week), or
- deplete, by exploiting it thoroughly (e.g. by daily fishing).

In terms of economic benefits in the long term (measured in conventional units), the possible results of competition are summarized in the *payoff* matrix displayed in Fig. 8.2.

If both exploiters adopt a conservative strategy, the fish population is kept in good condition and both A and B get 3 units of economic benefit (suitably measured as the difference between revenues from the catch and fishing costs) for a total of 6. If one of the two competitors adopts a depletion strategy, the population is not in very good conditions and, in the long term, provides a lower catch to both with a resulting total economic benefit of 5 units. However, the greedy exploiter gets more than the

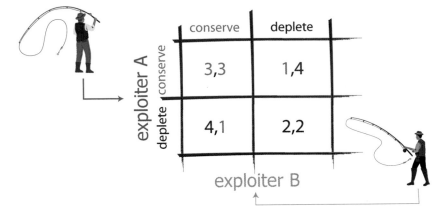

Fig. 8.2 Economic benefits to exploiters A and B deriving from employing four possible pairs of strategies "conserve" and "deplete" (see text). The numbers in each cell display the benefits accruing to A and B, respectively

conservative fisherman: 4 units. If both fishermen adopt a depletion strategy, the population is overexploited, its growth rate is greatly reduced and the catch is even lower. Thus the total economic benefit is 4 units, which are evenly shared between the two competitors.

It is instructive to wonder whether there is a pair of strategies that can actually establish in the long run. To answer this question, one can investigate which decision each of the two fishermen should take in the absence of any information on the strategy chosen by the competitor. Let's start from the best decision for exploiter A. If B decides to use a conservative strategy (first column of the table shown in Fig. 8.2) A has an incentive to use a depletion strategy. In this way, in fact, the fisherman would gain 4 units (first value of the pair of numbers in the second row and first column of the table) instead of 3 (same, but in the first row). If, on the contrary, B decided for full exploitation (second column), A would still take advantage from depleting the resource, because the profit would be 2 units (first value of the pair in the second column and second row) instead of 1 (same, but in the first row). Therefore, A would anyway decide to use a non-conservative strategy. The same reasoning can actually be made by fisherman B if any *a priori* information on A's strategy is not available. Given the symmetry of the *payoff* matrix, the same result is obtained that B should also fully exploit the resource. The conclusion is that, in the long run, a situation would establish in which both competing fishermen would adopt a non-conservative strategy, thus exploiting the resource in the worst possible way.

This dilemma highlights how the anarchy of completely open access leads to inevitable consequences, which are unpleasant under any point of view. In fact, from the viewpoint of species conservation, overexploitation endangers the viability of the populations being harvested; from the economic viewpoint, on the other hand, each of the exploiters could get more at a lower cost (implied by a lower exploitation effort), provided an agreement is reached by the competitors. In fact, if both competitors adopted a conservative strategy, the individual profit would be 3 instead of 2 units. Therefore, even economists have a fairly widespread belief that some kind of regulation in the exploitation of renewable resources should be introduced. This need is now accounted for not only at the level of individual countries, as evidenced by the various national laws on hunting, fishing and the management of forest resources, but also in the relationships between different countries, as evidenced by the international conventions (such as CITES, Convention on International Trade in Endangered Species of Wild Fauna and Flora) and various international regulatory commissions that manage the exploitation of biological resources whose habitat extends to the international geographic scale (for example, *InterAmerican Tropical Tuna Commission* or *International Whaling Commission*).

Of course, the regulatory policies that can be implemented depend on both the conditions in which exploiters are actually operating and the goals of regulators. These are the most various and do depend on personal beliefs, moral values or political commitments. The objectives can, at least from a historical point of view, be catalogued into three different categories. The first kind of regulatory policies, favourably considered by the traditional biological school, is based on the idea that the amount of harvested biomass should be maximized in the long term (the so-called concept of maximum sustainable yield, MSY). The second kind of goals, supported by resource economists (who consider the harvest of renewable resources as one of the many economic activities that can be entertained), is the maximization of the profitability by means of regulatory measures that are socially acceptable. The third kind of targets is that of species conservation and is mainly aimed at minimizing the risk of extinction of a population or limiting the deterioration of natural habitats, so as to best protect ecosystems. The discussion in the following sections provides consideration to the three points of view, which are all considered worthy of attention. It is to be remarked that even if one of the goals is considered of prominent importance, the other two can be taken into account by implementing suitable constraints. For instance, a regulatory agency might decide that the main goal is that of maximizing the net economic benefit with the constraint that the population extinction risk does not exceed a certain level.

The control policies with which decision makers can *de facto* implement the objectives described above are the most various and can be both technical and administrative. They range from the imposition of quotas that limit the harvest of animal or plant biomass, to the limitation of the hunting or fishing season, to the concession of licenses, to the imposition of taxes, to the preservation of a minimum viable population. All these policies will be analysed in this chapter as regards both their actual implementation and their consistency with the objectives to be achieved.

Some of the regulations are mutually alternative, but many can be used simultaneously. For instance, a decision maker can contemplate both a policy of hunting licenses and a policy that every year sets the maximum number of animals of a given species that can be caught. A more detailed description of the different methods for controlling the exploitation of a population, and their advantages and disadvantages, will be provided in the next sections, following in substance the classification given by Clark (1976). We must anyway remark that, precisely because of the key role that open access plays in the exploitation of living resources, we can distinguish between the different regulatory instruments on the basis of whether they affect or not the exclusiveness of harvesting rights. The more traditional management tools are substantially *non-exclusive*, i.e. they do not in principle preclude anyone from the right of hunting or fishing or lumbering. Among the main ones we cite:

1. *Restrictions to the harvesting time and place*: seasons in which hunting or fishing is allowed, areas within which hunters and fishermen can or cannot operate.
2. *Restrictions to hunting and fishing methods*: characteristics of the hunting or fishing gear (type of weapon or gear, use of dogs or lures, fishing-net mesh size, number of hooks and lines, etc.), allowed characteristics of fishing vessels (size, tonnage, horse power, equipment).
3. *Total quotas*: maximum number of animals of a given species that can be fished or hunted in a given area and in a certain year, maximum biomass that can be harvested in a given area and year; hunting or fishing or lumbering is not allowed when the quota is reached.
4. *Restrictions to the catch*: restrictions to the species and/or size and/or age and/or sex of the organisms present in the harvest, restrictions to by-catches and discard.

However, there are other regulatory tools that are exclusive, i.e. that exclude a certain number of hunters or fishermen from the right of harvesting. Among these, particularly relevant methods are

1. *licensing*: hunting or fishing is permitted each year only for a limited number of operators who hold a harvesting license;
2. *allocated quotas*: catch quotas are allocated to each holder of a license (either an individual or a company).

An interesting tool that does not belong to any of the previous categories is that of *financial disincentives*, namely taxes or royalties on the catch or the fishing effort or the finished product resulting from the harvesting activity. They represent an indirect way of controlling the exploitation of renewable resources.

8.2   Dynamics of a Harvested Population

Before analysing the pros and cons of the various possible management policies, it is fundamental to study the effects of human harvesting on the dynamics of an animal or plant population. To this purpose consider the simple case of a population with continuous reproduction in which the effects of age structure and the interactions with other populations are negligible compared to the intraspecific phenomena leading to density dependence. Denote by $h(t)$ the current harvest rate expressed, for example, as the number of organisms (or their biomass or their density) that are hunted, fished or generally harvested in the unit time. The population dynamics is then given by

$$\dot{N} = NR(N) - h(t) \tag{8.1}$$

where N is the abundance or biomass or density of the population, $R(N)$ is the per capita (or per unit biomass or per unit density) rate of growth in the absence of exploitation. As a first step it is interesting to analyse the effect of a constant rate of harvest over time. Therefore assume

$$h(t) = \bar{h} = \text{constant}$$

namely

$$\dot{N} = NR(N) - \bar{h}. \tag{8.2}$$

It is very simple to analyse the outcome of exploitation from a graphical point of view, as shown in Fig. 8.3. If $0 < \bar{h} < \dot{N}_{max}$, constant harvesting creates steady states ($\dot{N} = 0$) other than natural equilibria. If $\bar{h}$ is larger than the maximum value of the population growth rate of $NR(N)$ (which occurs in correspondence to $\bar{N}_{max}$) there are no steady states and the population is rapidly driven to extinction. Otherwise there exist two equilibrium populations ($N = \bar{N}_-$ and $N = \bar{N}_+$). As $\dot{N} < 0$ if $N < \bar{N}_-$ or $N > \bar{N}_+$, and $\dot{N} > 0$ if $\bar{N}_- < N < \bar{N}_+$, it is simple to derive that $\bar{N}_-$ is unstable and $\bar{N}_+$ is stable.

Thus we can conclude that, if by some random event (a year with a particularly cold winter or an invasion of warm tropical waters such as the one occurred in the Peruvian *anchoveta* (*Engraulis ringens*) collapse, which will be described in the sequel) the population fell below the threshold $\bar{N}_-$, it would be doomed to extinction

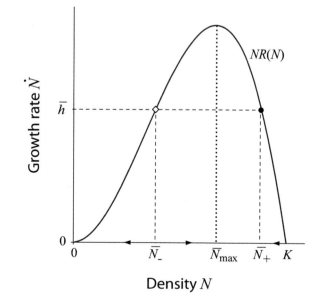

Fig. 8.3 In a population with growth rate $NR(N)$ (which achieves its maximum at $\bar{N}_{max}$) and carrying capacity K, a constant harvest rate $\bar{h}$ creates two steady states $\bar{N}_-$ and $\bar{N}_+$. The first is unstable, the second is stable

in case the exploiters persisted in applying the same rate of harvest $\bar{h}$. Note that this possibility is not unrealistic because in many cases (typically the exploitation of fish stocks) the abundance $N(t)$ of the population is not generally known. This analysis, though simplistic, is not useless because, as we will see later, one of the regulation measures (that of total catch quotas) acts directly on the harvesting rate.

An example of catastrophe like the one just analysed from the formal viewpoint and related to the fishing industry is the harvest of the Peruvian *anchoveta*, a very big anchovy stock (*Engraulis ringens*). This fishery developed in the 1960s s and became, in terms of caught biomass, the largest in the world. Biologists estimated as optimal a catch quota in the order of 10 million tonnes per year, which was reached in 1967 as shown in Table 8.1. However the fishing capacity in terms of both fishing fleet and processing plants (most fresh product is converted into fish-meal) developed up to a level far higher than what is necessary to harvest the optimal catch. In 1971 (see again the Table 8.1) the quota was captured in less than three months with the result that, for the rest of the year, fishing vessels and plants remained idle.

Between 1972 and 1973 came the collapse. It looks like (Glantz 1979) the reproduction of anchovy was heavily disturbed in 1972 because of a strong *El Niño* (which is a warm ocean current of variable intensity that develops after late December along

Table 8.1 The development of the Peruvian *anchoveta* fishery between 1959 and 1978. Missing data are not available

Year	Number of vessels	Number of fishing days	Catch (10^6 tonnes)
1959	414	294	1.91
1960	667	279	2.93
1961	756	298	4.58
1962	1069	294	6.27
1963	1655	269	6.42
1964	1744	297	8.86
1965	1623	265	7.23
1966	1650	190	8.53
1967	1569	170	9.82
1968	1490	167	10.26
1969	1455	162	8.96
1970	1499	180	12.27
1971	1473	89	10.28
1972	1399	89	4.45
1973	1256	27	1.78
1974	–	–	4.00
1976	–	–	4.30
1977	–	–	0.80
1978	–	42	0.50

the coast of Ecuador and Peru and sometimes causes catastrophic weather conditions). In particular, during *El Niño* the normal up-welling that brings nutrients to surface waters is weakened, primary production is reduced and so is secondary production too. The 1972 event caused a sudden decline in stock abundance. At first daily catches were still high because the fleet, much more skilful than it was in the past, was able to concentrate on what remained of the fish stock. When the enormity of the collapse became apparent, the *anchoveta* stock had been virtually wiped out. If one considers that the export of fish-meal obtained through the fishery was the second most important source of foreign currency for Peru, one can understand the dramatic impact this catastrophe had on the economy of this South American country. More recent research (Klyashtorin 2001) ascertained that the collapse of Peruvian anchoveta is not exclusively due to an erroneous management of the allowed catch, but also to a number of contributing factors on the top of which excessive harvesting became the compromising factor.

The analysis so far illustrated has the drawback of introducing the harvest rate h as the main decision variable of the renewable resource exploiters. In fact, in most cases the harvest rate does not constitute a decision variable, namely a variable that can independently be decided upon by the economic operators who exploit a natural population. Indeed, h does depend on the abundance of the population: in fact, given a certain level of human commitment to exploiting a resource, the resulting harvest rate is larger if the resource is abundant. For this reason, it is necessary to introduce the concept of effort. With the term *effort* we mean a suitable measure, in appropriate units, of the amount of capital and labour committed to exploiting some renewable resource. Examples of effort indices are: the daily number of vessels of a flotilla engaged in harvesting a fish stock; the number of licensed hunters multiplied by the fraction of those who actually search for animals to be killed; the number of traps or fishing gears employed in a given region; the number of lumberjacks employed in forestry activities. If one considers that the exploitation activity is actually predation by humans, the effort is proportional to the number and degree of activity of human predators. Therefore if we denote by $E(t)$ the harvesting effort at time t, the rate of resource harvest is linked to both $E(t)$ and $N(t)$ by expressions that are similar to those employed to describe predation.

More precisely, if one assumes that the resource is uniformly distributed in space, the activity of resource searching is random and there is no limit to the harvest rate obtained by a unit of effort, then one can stipulate that

$$h(t) = qE(t)N(t) \qquad (8.3)$$

where q is a parameter called *catchability coefficient* which translates how effective one effort unit is in harvesting the resource. Given a certain population, the value of q depends of course on the technology used for capturing individuals of that population and on the ability of the resource exploiters. Usually the catchability coefficient is assumed to be constant, although it certainly varies over time, especially in the mid to long term. In fact, it is conceivable that q increases because hunting, lumbering and fishing techniques become more and more sophisticated and destructive.

If, in analogy to what was done for predation, we assumed that there is a limit to the amount of resource that can be harvested in the unit time by a single unit of effort (for example, due to saturation of the fishing gear), then the expression for h would be of the type:

$$h = qE\Phi(N)$$

where $\Phi(N)$ is nothing but the functional response of the predator and thus it is a function usually growing up to a saturation level.

Now let us confine ourselves to the simple case given by the relation (8.3) and initially analyse the consequences of keeping the effort constant over time, i.e.

$$E(t) = \bar{E} = \text{constant} .$$

The population dynamics is now described by the equation

$$\dot{N} = N(R(N) - q\bar{E}). \tag{8.4}$$

The steady states of equation (8.4) can be found in a rather simple way, for example graphically (see Fig. 8.4).

In the case of pure compensation, namely when the per capita growth rate is steadily decreasing with N, there exist two equilibria ($N = 0$ and $N = \bar{N}$) if the effort $\bar{E}$ is not too high, while there exists only one ($N = 0$) if

$$\bar{E} > \frac{R(0)}{q}.$$

In the first case, it is easy to establish that $\bar{N}$ is stable, while $N = 0$ is unstable, because $\dot{N} < 0$ if $N > \bar{N}$, $\dot{N} > 0$ if $0 < N < \bar{N}$. In the second case the extinction

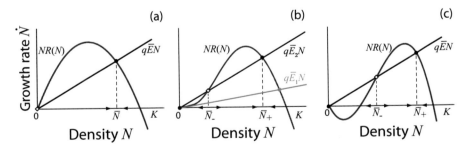

Fig. 8.4 A constant effort $\bar{E}$ gives rise to different steady state situations according to population demography. **a** Pure compensation: there exists only one stable equilibrium $\bar{N}$; **b** Non-critical depensation: for low effort ($\bar{E}_1$) there exists only one stable equilibrium $\bar{N}$, while high effort ($\bar{E}_2$) gives rise to three equilibria: one is unstable ($\bar{N}_-$), and two are stable $\bar{N}_0$ and $\bar{N}_+$; **c** Critical depensation: there exist three equilibria for any effort: one is unstable ($\bar{N}_-$), and two are stable ($\bar{N}_0$ and $\bar{N}_+$)

equilibrium $N = 0$ is instead stable because the harvest rate qEN is greater than $NR(N)$ for any value of $N > 0$.

The result obtained through the model (8.4) is thus qualitatively different from that obtained with the model (8.2) in which the harvest rate was constant. In fact, if effort is limited within reasonable levels, there is always a stable non-trivial equilibrium and there is no threshold below which the population is doomed to extinction. Any fortuitous accident that may lead the population below the level $\bar{N}$ can in fact be absorbed over time. This is basically due to the fact that the harvest rate, being now not fixed but proportional to the population size $(N(t))$, is automatically reduced when N decreases.

It should however be noted that if there is evidence of depensation phenomena (or Allee effect), be it critical or non-critical, such as those described in Sect. 2.4 of Chap. 2 the situation can be quite dangerous even if a policy of constant effort is used. This is demonstrated in panels (b) and (c) of Fig. 8.4. If depensation is critical, there exist (for not too high efforts $\bar{E}$) three equilibrium states: $\bar{N}_0 = 0$, $\bar{N}_-$ and $\bar{N}_+$. Extinction $\bar{N}_0$ and $\bar{N}_+$ are stable, while $\bar{N}_-$ is unstable, because

$$\dot{N} < 0 \text{ if } 0 < N < \bar{N}_- \text{ or } N > \bar{N}_+$$
$$\dot{N} > 0 \text{ if } \bar{N}_- < N < \bar{N}_+.$$

Therefore, if for any reason the population size falls below the threshold $\bar{N}_-$ the population is doomed to extinction.

If depensation is not critical and exploitation efforts are not too high (case $\bar{E} = \bar{E}_1$ of panel b in Fig. 8.4), there is no threshold phenomenon, while there is a threshold abundance for intermediate efforts (case $E = \bar{E}_2$ of Fig. 8.4b). As usual, extinction ensues from excessively high values of effort.

A quite similar analysis to the one just illustrated can be conducted in the case of populations for which reproduction is concentrated in a restricted season of the year. They are thus better described by difference equations (as those shown in Sect. 2.2). Assume that a population of this type is exploited during a time period that does not coincide with that of reproduction. We will refer to this time period as fishing or hunting or more generally harvesting season. The time sequence of events that mark the dynamics of an exploited population from one year to the next is shown schematically in Fig. 8.5. In the scheme we indicate with P_k the so-called *parental stock*, namely the abundance or density of reproductive adults during the reproductive generation k. This parental stock will produce Y_k youngsters. The juvenile and parental stocks will then face a period during which they will be subjected to death risks due to causes other than exploitation by man. Usually, harvesting is forbidden during the reproductive season and the periods of high mortality. At the beginning of the hunting, fishing or harvesting season, there will eventually be a number of individuals R_k on which harvesting can be exerted. The population size R_k is called *recruitment*. A portion H_k of the recruitment is then harvested and the remainder is the parental stock P_{k+1} of the next generation $k + 1$. The precise kind of relationships between the involved variables $(P_k, Y_k, R_k, H_k, P_{k+1})$ change of course from

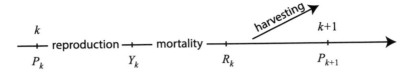

Fig. 8.5 Temporal sequence of relevant life events in a harvested population with concentrated reproduction. k is the generation, while P, Y, and R, are the parental stock, the youngsters and the recruitment, respectively

population to population. The effects of density dependence can indeed be operating only in certain phases of the life cycle, or in all. If, for instance, reproduction and adult mortality are not dependent on density, but juvenile mortality is dependent, then

$$Y_k = f P_k$$
$$R_k = G(Y_k) + \sigma P_k \tag{8.5}$$

where f represents the average fertility of each adult and σ is the adult survival. In some cases, σ can be zero, for instance in annual plants or in those animal species (such as the Pacific salmon, genus *Oncorhynchus*) that die immediately after reproducing (so-called *semelparous* species). The function $G(Y_k)$ links recruitment to the abundance of juveniles and is generally non-linear because of dependence on density. Combining expressions (8.5) we finally obtain

$$R_k = G(f P_k) + \sigma P_k = F(P_k) \tag{8.6}$$

The function $F(P_k)$ is called *stock-recruitment* function.

Note, however, that it is not necessary to postulate that density dependence is restricted to juvenile mortality in order to obtain a relation of the kind (8.6). More generally one can assume

$$Y_k = N(P_k) R_k = G(Y_k) + S(P_k)$$

and obtain

$$R_k = G(N(P_k)) + S(P_k) = F(P_k).$$

For example, Lett et al. (1981), using data collected for several decades from a population of harp seals (*Pagophilus groenlandicus*), have shown that density dependence affects reproduction (see panel a in Fig. 8.6), although quite mildly, and mortality of less-than-one-year-old pups (Panel b in Fig. 8.6). It looks like adult survival (estimated at about 90% yearly) is not highly dependent on density.

To link the parental stocks between generations, one must of course take into account harvest H_k. By assuming that the natural mortality of recruitment is negligible during the harvesting period, it turns out that

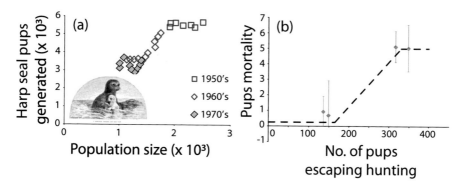

Fig. 8.6 a Number of harp seal (*Pagophilus groenlandicus*) pups generated by one-year-old or older seals as a function of the population size. Inset is an ancient plate showing a harp seal mother with her baby. **b** Pup mortality as a function of the number of pups that have escaped hunting. Vertical bars indicate two standard deviations. Redrawn after Lett et al. (1981)

$$P_{k+1} = F(P_k) - H_k \tag{8.7}$$

An alternative way of writing (8.7) that highlights the link of recruitments, instead of parental stocks, between generations is the following

$$R_{k+1} = F(P_{k+1}) = F(R_k - H_k). \tag{8.8}$$

To analyze the different effects of harvesting on the population dynamics one must consider that, as H_k is total harvest, not harvest rate, the obvious constraint must be satisfied that an exploiter cannot fish or hunt more than what is available at the beginning of the harvesting season, namely that

$$H_k \le R_k.$$

In this discrete-time case too it is instructive to examine the effect of a constant harvest, or more precisely

$$H_k = \begin{cases} \bar{H} & \text{if } R_k > \bar{H} \\ R_k & \text{if } R_k < \bar{H} \end{cases}$$

because total harvest cannot anyway exceed recruitment. The population dynamics can be graphically analysed in a simple way by using a cobweb plot for discrete-time population models. The relationship between parental stocks from one generation to the next is represented in Fig. 8.7 by the black curve. There are three intersections between the curve and the 45° line $P_{k+1} = P_k$, which represent the resulting equilibria, namely $P_k = 0$, $P_k = \bar{P}_-$, $P_k = \bar{P}_+$. Similarly to the continuous case, $\bar{P}_-$ is unstable and acts like a threshold below which the population is doomed to extinction.

Fig. 8.7 A constant harvest $\bar{H}$ alters the original stock-recruitment curve $F(P)$ (in grey) giving rise to the new curve linking the parental stocks (in blue). Therefore, the equilibria are P_- (unstable) and P_+ (stable). The symbol K, as usual, indicates the carrying capacity in natural conditions

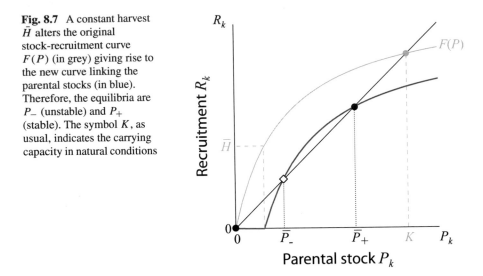

Even in the case of discrete generations one can introduce the concept of effort and derive the relationship between effort and catch. To that end we must analyse in more detail the course of events during the harvesting season. We denote by 0 and T, respectively, the initial and the final times of the season and by $N(t)$ and $E(t)$, respectively, the population size and the effort exerted at the current time t during the season. If we assume that the harvest rate is related to the effort and the population size in accordance with the simple equation (8.3), the decrease of N in the unit time is given by

$$\dot{N} = -qEN.$$

By integrating this equation one obtains

$$N(T) = N(0) \exp\left(-q \int_0^T E(t)dt\right)$$

On the other hand, $N(0)$ is nothing but the recruitment R_k and $N(T)$ is the parental stock P_{k+1}. Instead, the integral $\int_0^T E(t)dt$ represents the cumulated effort during the kth harvesting season. If we introduce the duration T_k of the harvesting season and the average effort E_k in year k, then $T_k E_k = \int_0^T E(t)dt$. Therefore

$$P_{k+1} = R_k \exp\left(-q T_k E_k\right)$$

from which it finally turns out

$$H_k = R_k - P_{k+1} = R_k \left(1 - \exp\left(-q T_k E_k\right)\right). \tag{8.9}$$

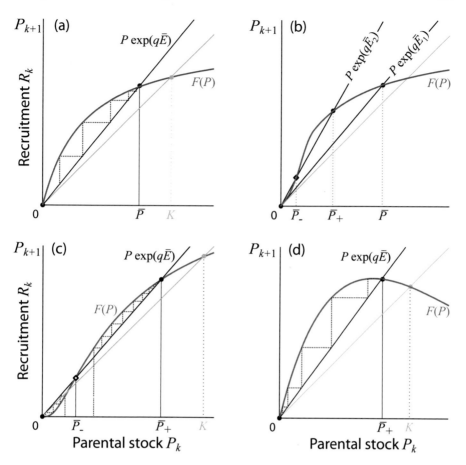

Fig. 8.8 A constant cumulated harvesting effort $\bar{E}$ exerted on a population with concentrated reproduction can give rise to different steady state solutions according to the kind of population dynamics in the absence of exploitation. Cobweb plots and equilibria are shown. **a** Compensation: there exists only one equilibrium $\bar{P}$ which is stable; **b** Non-critical depensation: for low effort ($\bar{E}_1$) there exists only one stable equilibrium $\bar{P}$, while for higher effort ($\bar{E}_2$) there exist three equilibria one of which is unstable ($\bar{P}_-$), the other two ($\bar{P}_0 = 0$ and $\bar{P}_+ > \bar{P}_-$) are stable; **c** Critical depensation: under any case (that is, independently of the exerted effort) there exists an unstable equilibrium $\bar{P}_-$. Like in **b**, equilibria $\bar{P}_0 = 0$ and $\bar{P}_+$ are stable; **d** Overcompensation: there exists only one equilibrium $\bar{P}$, not necessarily stable like the one in the case shown in the figure

This expression links harvest, recruitment and cumulated effort. Note that, as effort increases, catch tends to equal recruitment. When there is inhomogeneity in the spatial distribution of the resource, absence of randomness in its search and gear saturation, the specific expression (8.9) should be replaced by other more complicated formulas.

Let us stick to the simpler hypothesis and explore the consequences of a total cumulated effort which is kept constant in time, i.e.

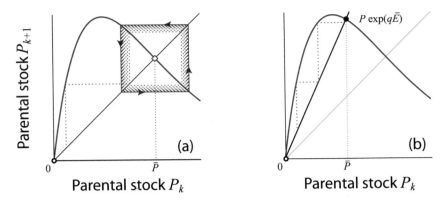

Fig. 8.9 A sufficiently high effort can lead to the elimination of the oscillations (**a**) possibly occurring in the natural situation (**b**)

$$E_k T_k = \bar{E} = \text{ constant.}$$

The resource dynamics is thus given by

$$P_{k+1} = F(P_k) \exp\left(-q E_k T_k\right) = F(P_k) \exp\left(-q\bar{E}\right).$$

The graphical analysis is similar to the one utilized for the dynamics of populations that are not subjected to exploitation, provided one replaces the 45° line with the straight line having a slope equal to $\exp\left(q\bar{E}\right)$. The results of the analysis are shown in the panels of Fig. 8.8 which displays the cases of compensation (panel a), non-critical depensation (b), critical depensation (c) and overcompensation (d). From the qualitative point of view the results are equivalent to those already discussed for continuous-time populations, with the only exception of overcompensation (panel d). The application of a sufficiently high effort, in fact, can lead to the elimination of potential fluctuations in the population's natural dynamics (that is in the absence of exploitation), as outlined in Fig. 8.9.

8.3   Effects of Some Regulation Policies

The dynamical analysis carried out in the previous section allows us to investigate the effects of the most commonly implemented policies aimed at controlling exploitation activities. We will discuss the relative strengths and weaknesses with particular regard to the influence of these policies on the basic problem of the time evolution of exploited populations, without addressing for the moment more complex issues such as the possible search for policies that allow the optimal achievement of some given goals.

The simplest, and possibly most used, form of regulation implemented in those cases in which an institution (state agency, international commission, etc.) is given the power to monitor the activities of resource harvesting, is the so-called *total quota system*. It consists in allowing the exploiters to harvest no more than a fixed number or a given biomass (the "quota", as a matter of fact) of animals or plants in a certain time period. This method was used, for example, by the *International Whaling Commission (IWC)* to regulate whale hunting before 1974.

Although the quota may be changed between years and in principle is only an upper bound to the actual harvest, in practice the quota is rarely changed and is very often equal to the harvest, because access to the resource is open to any number of exploiters. The effect of a harvest rate that is kept constant over time has been already analysed in the previous section. We showed that it makes the exploited population quite vulnerable to environmental vagaries. Therefore, imposing a fixed, invariable quota over time is decidedly deleterious because it is likely to turn a bad year into an irreversible catastrophe. This criticism is usually countered by arguing that quotas can be changed from time to time; but what has been in fact occurring historically is that a quota is fixed once and for all on the basis of what seems a suitable criterion and is changed only after the decision makers realize that something unpleasant has occurred. This recognition is often late, because of the inherent and frequent difficulties in evaluating the abundance of a population.

Significant in this regard are the Tables 8.2 and 8.3 (both after Watt 1968). Table 8.2 reports world catches of whales in the second half of the past century and the relevant quotas imposed by IWC in terms of blue whale units. In fact, quotas were established every year in an aggregate manner for the different species, by stipulating the equivalence of a blue whale (*Balaenoptera musculus*) with two fin whales (*Balaenoptera physalus*) and six sei whales (*Balaenoptera borealis*). Note that the quota remained

Table 8.2 Total world catches of whales and relevant IWC quotas in the period following the second world war, expressed as blue whale units

Year	Total catch	Quota
1947–1948	16,364	16,000
1952–1953	14,866	16,000
1957–1958	14,850	14,500
1958–1959	15,300	15,000
1959–1960	15,511	15,000
1960–1961	16,433	–
1961–1962	15,252	–
1962–1963	11,306	15,000
1963–1964	8,429	10,000
1964–1965	6,987	8,000
1965–1966	4,091	4,500
1966–1967	–	3,500

Table 8.3 Change of species composition in the world whale catch as referenced in Table 8.2

Year	Catch of all whale species	Catch of blue whales
1937–1938	54,902	15,035
1947–1948	43,431	7,157
1950–1951	55,795	7,278
1951–1952	49,832	5,436
1952–1953	45,009	4,218
1953–1954	53,642	3,009
1954–1955	55,074	2,495
1955–1956	58,126	1,987
1956–1957	59,056	1,775
1957–1958	64,586	1,995
1958–1959	64,489	1,442
1959–1960	63,717	1,465
1960–1961	65,811	1,987
1961–1962	66,026	1,255
1962–1963	63,579	1,429
1963–1964	63,001	372
1964–1965	–	20
1965–1966	–	

virtually constant from 1947 to 1962; also, the capture is constant until 1961, but only at first glance because in reality the quota, being expressed in aggregate units, does not reflect the fact that the big blue whale is driven to extinction (as highlighted in Table 8.3). Since 1963, the quota was progressively lowered and interestingly enough was systematically higher than the catch that actually materialized. This indicates that the control method adopted by IWC was actually regulating nothing, indeed it was practically in tow of events.

Therefore, the quota system requires constant surveillance of the abundance of exploited populations in order to verify that it does not drop below a safe level. However this kind of surveillance is usually quite expensive because it requires the collection of many data by utilizing e.g. capture, mark, and recapture techniques or aerial surveys or sonar stock assessments.

A second form of regulation is *effort limitation*. This policy was for example used for the management of Pacific salmon (genus *Oncorhynchus*) fishing. As made clear in the previous section, with regard to discrete-time demography, the limitation of the total cumulated effort is achieved by both regulating the duration of the harvesting season and the average effort (e.g. the average number of operating fishing vessels). Even in this case the allowed effort can vary over time and is an upper bound (maximum allowed effort), but in practice it is fixed on the basis of appropriate criteria and is changed only in the face of unexpected events. However, as evidenced in the previous section, the imposition of a constant effort does not have the deleterious effects of fixed quota systems. If there are no depensation phenomena, in fact, this method ensures a sort of automatic regulation of the harvest rate without the need to set up an expensive system of population abundance assessment (recall Figs. 8.4a and 8.8). Of course, being able to impose that effort is actually lower than a constant upper bound can be a problem. But this problem is no more difficult than being able to impose that a certain harvest rate is not exceeded. The instruments used to limit effort are quite varied and range from granting a number of licenses to hunt or fish, to limiting the duration of the season in which it is allowed to exploit a renewable resource. Strengths and weaknesses of these instruments will be discussed in Sect. 8.5.

A real drawback is instead represented by the fact that the exploited population dynamics is not sensitive to effort only but to the product of the catchability coefficient and the effort, namely to qE. In many cases it is not reasonable to assume that q is invariant in time, because technology (or even just the harvesting ability of exploiters) tends to become more sophisticated, thus increasing q. Therefore, limiting effort can not be a sufficient measure, unless this policy is coupled to measures that control the technology used; these are in fact many times enforced by specifying the characteristics of fishing gears and hunting weapons (for example by ruling the minimum mesh size of fishing nets).

Finally, another aspect highlighted by the analysis of constant effort policies is the vulnerability of those populations that, due to their social structure or to the considerable predation pressure at low density, are subject to depensation. The greatest attention must thus be paid to managing populations with this kind of demography.

When species are better described by time-discrete dynamics and are thus subjected to harvesting during a period that does not overlap with the reproductive season, a further policy deserves to be mentioned. Although such a policy is not easy to implement, it provides considerable advantages from an environmental viewpoint. It consists of keeping the spawning stock as constant as possible. The spawning stock originates from the stock that survives harvesting and is therefore termed escapement. That's why this policy is often called *constant escapement*. In other words, the total harvest H_k is determined each year according to the following rule

$$H_k = \begin{cases} R_k - \bar{P} & \text{if } R_k \geq \bar{P} \\ 0 & \text{if } R_k < \bar{P} \end{cases}$$

where $\bar{P}$ is the desired escapement. All the biomass exceeding the escapement $\bar{P}$ is thus harvested and the remainder is available for reproduction; if the recruitment is smaller than the established escapement, harvesting is forbidden. The population dynamics is in this case given by the equation

$$P_{k+1} = \begin{cases} \bar{P} & \text{if } F(P_k) \geq \bar{P} \\ F(P_k) & \text{if } F(P_k) < \bar{P} \end{cases} \tag{8.10}$$

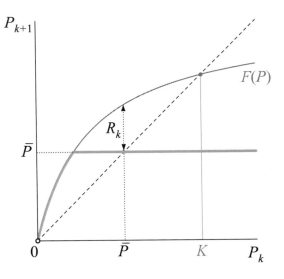

Fig. 8.10 The constant escapement policy ($\bar{P}$, bold cyan curve) is applied to a discrete-time population with stock-recruitment function $F(P)$ (pink curve)

Graphically, the dynamics is represented in Fig. 8.10. It is easy to conclude that, unless the population is characterized by critical depensation, it converges to a stable equilibrium which obviously coincides with $\bar{P}$.

This regulation system is thus very attractive from the point of view of species preservation. It ensures the persistence of the population even if this displays non-critical depensation, thus virtually averting any risk of extinction. Constant effort policies, instead, may imply local stability of extinction under similar conditions.

Not surprisingly, this method has been used to regulate Pacific salmon fishing in Canada, a country that is extremely sensitive to the problems of biological conservation. The main disadvantage of the *constant escapement* system is that it requires a highly sophisticated apparatus for controlling the population size, more specifically the recruitment. In some (rare) cases salmon are counted one by one while they go upstream, as is done in some of the North American rivers. Alternatively, other sampling methods can be used, which may also be quite expensive, especially if the species to be exploited is not concentrated in a restricted area (as Pacific salmon is, because this fish returns to the estuary of the river from which the stock originates) but is spread over a large area. A second disadvantage of this policy is that in "bad years" (recruitment R_k smaller than $\bar{P}$) no harvest is allowed, unlike the case of a constant effort policy, according to which a fixed proportion of the recruitment is taken under any condition. This is not always and everywhere socially acceptable. It requires that the competent government does organize, in addition to a complicated and expensive surveillance system of the stock, an equally expensive system of subsidies to compensate for the lost harvest.

8.4 The Problem of Maximum Sustainable Yield

The simplest approach to the problem of rationally planning the exploitation of biological resources is to search for the policy that allows, at equilibrium, the maximum biomass harvest and thus the maximum yield. The optimization of harvest according to this criterion was very popular in the 1950s and 1960s and is still very much used at least as a benchmark in many fishing and hunting management textbooks. This viewpoint considers the renewable resource like a farm product or a centrally managed crop which must be dimensioned so as to ensure the maximum production in perpetuity. As a result, the optimal policy would be the one that leads to the famous concept of "Maximum sustainable yield" (the international acronym is *MSY*). By the term "sustainable" we mean that the population must be at a stable equilibrium in the long run. Although the concept started being criticized more than forty years ago (Larkin 1977), it must be recognized that it is one the first examples of the sustainability concept (Gatto 1995). This section analyses how to achieve the objective of maximum sustainable yield by the control policies illustrated in the previous section. In other words, given a certain class of policies, we will find the one that is optimal according to the MSY objective.

Consider first the total quota system. In this case, the quota in a certain time interval coincides with the production in the same interval. Therefore, if one wants to maximize the sustainable yield one must impose the maximum quota, constant in time. With reference to model (8.2) and Fig. 8.3, then the decision maker must impose a harvest rate $\bar{h}$ equal to the maximum growth rate, namely

$$\bar{h}_{\max} = N_{\max} R(N_{\max})$$

With the optimal quota policy, however, the two equilibria $\bar{N}_-$ (unstable) and $\bar{N}_+$ (stable) displayed in Fig. 8.3 merge in the unique equilibrium $N_{\max}$, to which the population size tends if the initial condition is larger than $N_{\max}$, but from which it is repelled toward extinction if the initial condition is smaller (see panel a of Fig. 8.11). Therefore, if the optimal quota policy is used in quite a short-sighted way with the aim of achieving the maximum sustainable yield, the slightest environmental disturbance that brings the population below the equilibrium $N_{\max}$ is likely to cause a possible extinction risk (if the same quota keeps being employed). To sum up, the combination of the quota policy with the objective of maximum sustainable yield is a highly dangerous mix and should be avoided carefully.

Suppose instead that a regulation system based on effort containment is used. What is the constant effort that should be employed to ensure the maximum sustainable biomass harvest? The problem is easily solved graphically in the case of both continuous reproduction and discrete-time demography. Let the dynamics be given by equation (8.1), namely

$$\dot{N} = NR(N) - qEN.$$

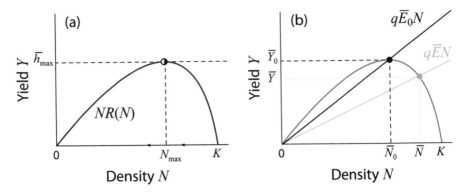

Fig. 8.11 **a** Effects of a quota h that achieves the maximum sustainable yield. As indicated by the arrows reported on the abscissa axis, the population size tends toward N_{max} for initial conditions larger than N_{max}, but decreases toward extinction for smaller ones. **b** If the population is exploited by means a constant effort $\bar{E}$, which implies an equilibrium population $\bar{N}$, a sustained yield $\bar{Y}$ is obtained. $\bar{E}_0$ is the effort that guarantees the maximum sustainable yield $\bar{Y}_0$, which corresponds to a density $\bar{N}_0$ obviously equal to N_{max}

Given a constant effort $E = \bar{E}$, the corresponding sustainable yield $\bar{Y}$ is given by

$$\bar{Y} = q\bar{E}\bar{N} = \bar{N}R(\bar{N}) \tag{8.11}$$

where $\bar{N}$ is the population size at equilibrium. Graphically, $\bar{Y}$ is the ordinate of the point of intersection between the population growth curve and the straight line with slope $q\bar{E}$ (see panel b in Fig. 8.11). It is easy to understand that the policy that ensures MSY is the one corresponding to the population $\bar{N}_0$ which guarantees the maximum population growth rate. The effort that must be employed is therefore $\bar{E}_0$.

In the case of pure compensation, to any effort $\bar{E}$ corresponds a single stable population equilibrium $\bar{N}$, and thus a single sustainable yield given by $q\bar{E}\bar{N}$. In other words, there exists a relationship between sustainable yield and effort, as shown in panel (a) of Fig. 8.12.

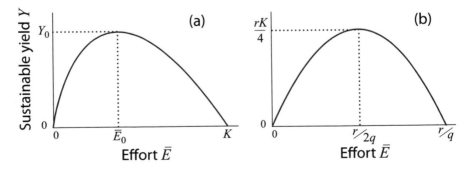

Fig. 8.12 Relationship between sustainable yield Y and effort $\bar{E}$ in a generic example of pure compensation **a** and in the case of logistic growth **b**, namely when the model is the one by Schaefer (8.12)

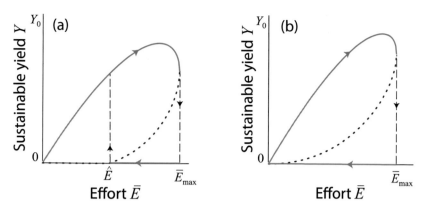

Fig. 8.13 Relationship between effort and sustainable yield in the cases of non-critical **a** and critical **b** depensation. Dashed curves indicate yields corresponding to unstable equilibria

Instead, if growth rate is characterized by depensation—whether critical or not—to a constant value effort may correspond more than one stable equilibrium (see Fig. 8.4, panels b and c) and, therefore, more than one sustainable yield. In such a case, there can exist a so-called *hysteresis effect*, as shown in Fig. 8.13. If effort is increased beyond the value $\bar{E}_{max}$, the sustainable yield falls to zero, since the two equilibria $\bar{N}_-$ and $\bar{N}_+$ collide through a bifurcation of catastrophic nature. The situation is reversible in the case of non-critical depensation (panel a of Fig. 8.13) provided effort is decreased below the level $\hat{E}$, otherwise the system would be trapped in the lower branch of the effort-yield curve. If the system is characterized by critical depensation, instead, 8.13b) the situation is irreversible, because extinction is anyway a stable steady state.

Historically, the first model used for calculating the maximum sustainable yield is that due to biologist M.B. Schaefer (1967). It simply posits logistic dynamics for the exploited species, namely

$$\dot{N} = rN \left(1 - \frac{N}{K} \right) - qEN. \tag{8.12}$$

In this case, for a given a constant effort $\bar{E}$, the corresponding equilibrium stock $\bar{N}$ is

$$\bar{N} = K \left(1 - \frac{q\bar{E}}{r} \right) \tag{8.13}$$

and the associated sustainable yield is given by

$$Y = q\bar{E}\bar{N} = qK\bar{E} \left(1 - \frac{q\bar{E}}{r} \right). \tag{8.14}$$

The conclusion is that, when viewed as a function of effort, the sustainable yield curve is shaped like a parabola with a maximum in correspondence of $\bar{E}_0 = \frac{r}{2q}$ (see panel b of Fig. 8.12). The maximum sustainable yield for the model (8.12) is $\bar{Y}_0 = \frac{rK}{4}$. Therefore, at MSY, the equilibrium stock is the one that provides the highest demographic growth rate, i.e. $\frac{K}{2}$.

For instance, Schaefer (1967) provides the following values of intrinsic growth rate and carrying capacity for the stock of *yellowfin* tuna (species *Thunnus albacares*) of the Eastern Pacific Ocean (see also Fig. 8.14)

$$r = 2.61 \text{ year}^{-1}$$
$$K = 1.34 \cdot 10^8 \text{ kg}.$$

He also evaluates the catchability coefficient to be given by

$$q = 3.8 \cdot 10^{-5} \text{ fishing days}^{-1}$$

so that the harvest rate $h = qEN$ is expressed in kilograms per year, provided the effort E is measured in number of fishing days per year and N in kilograms. The maximum sustainable yield should therefore be obtained with an effort equal to $\frac{r}{2q}$, that is 34,300 fishing days per year, and be equal to $\frac{rK}{4}$, that is 87,400 tonnes per year.

Similar considerations can be made when the dynamics is time-discrete and given, for example, by equations (8.7) and (8.9), that is to say by

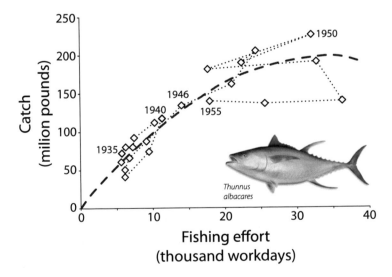

Fig. 8.14 Efforts and catches of *yellowfin* tuna (*Thunnus albacares*, inset) in eastern Pacific between 1934 and 1955. The parabola is the estimated relationship between effort and sustainable yield (Schaefer 1967)

$$P_{k+1} = F(P_k) \exp(-q E_k).$$

Given the constant effort $\bar{E}$, the corresponding parental stock at equilibrium is

$$\bar{P} = F(\bar{P}) \exp(-q\bar{E}).$$

Thus, sustainable yield Y is given by

$$Y = F(\bar{P})\left(1 - \exp(-q\bar{E})\right) = F(\bar{P}) - \bar{P} \tag{8.15}$$

which corresponds in Fig. 8.15 to the distance between the point of ordinate $F(\bar{P})$ belonging to the stock-recruitment curve and the point of ordinate $\bar{P}$ belonging to the 45° line. It is easy to demonstrate that the effort that allows MSY is thus the effort $\bar{E}_0$ that produces the equilibrium stock $\bar{P}_0$ characterized by

$$\left.\frac{dF}{dP}\right|_{\bar{P}_0} = 1$$

It is thus the stock at which the difference between $R_k = F(P_k)$ and P_k is maximum, namely at which the *surplus* biomass produced in one generation is maximum.

The relationships of the sustainable yield Y vs. the constant effort $\bar{E}$ are quite similar, from a qualitative viewpoint, to those of the time-continuous case as shown

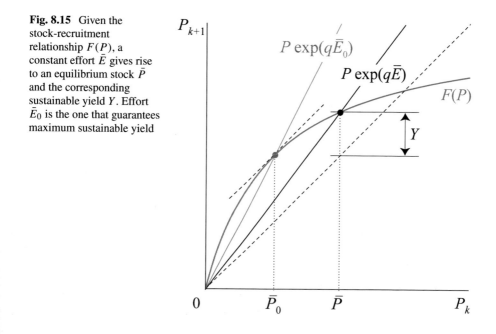

Fig. 8.15 Given the stock-recruitment relationship $F(P)$, a constant effort $\bar{E}$ gives rise to an equilibrium stock $\bar{P}$ and the corresponding sustainable yield Y. Effort $\bar{E}_0$ is the one that guarantees maximum sustainable yield

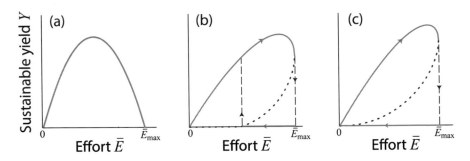

Fig. 8.16 Relationships of sustainable yield (solid green) vs. constant effort in the cases of: **a** compensation; **b** non-critical depensation and **c** critical depensation. Black dashed curves correspond to transitions from unstable equilibria (dotted red)

in Fig. 8.16a, b, c, respectively for the case of compensation, non-critical depensation and critical depensation.

It is also possible to solve the problem of maximum sustainable yield in the case in which a constant escapement policy is adopted. In this case the population dynamics is given by equation (8.10).

$$P_{k+1} = \begin{cases} \bar{P} & \text{if } F(P_k) \geq \bar{P} \\ F(P_k) & \text{if } F(P_k) < \bar{P} \end{cases}$$

Obviously, the steady state is $\bar{P}$ provided that $F(\bar{P}) > \bar{P}$. On the other hand, $\bar{P}$ is always chosen so as to satisfy $F(\bar{P}) > \bar{P}$, because the sustainable yield Y is given by

$$Y = F(\bar{P}) - \bar{P}$$

and of course it cannot be negative. From the graphical point of view, the situation is summarized in Fig. 8.17. The optimal policy is the constant escapement $\bar{P}$ that maximizes Y. By taking the derivative of Y with respect to $\bar{P}$ one can easily show that the optimal spawning stock $\bar{P}_O$ is the one that satisfies the relation

$$\left.\frac{dF}{dP}\right|_{\bar{P}_O} = 1.$$

Note that $\bar{P}_O$ coincides with the optimal spawning stock that corresponds to the optimal constant effort policy. Therefore the two policies, constant effort and constant escapement coincide at equilibrium, being indeed different from each other only with regard to the management that is applied under out-of-equilibrium conditions. This remark is actually very important, because a population is seldom at the deterministic equilibrium owing to the variability of external environmental conditions (environmental stochasticity). As an example, see Fig. 8.18 which reports the spawning stocks and the corresponding recruitments for the sockeye salmon (*Oncorhynchus nerka*)

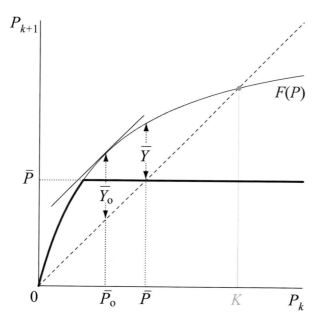

Fig. 8.17 The policy which allows for a constant escapement $\bar{P}$ guarantees, for the given stock-recruitemnt function $F(\bar{P})$, the sustainable yield Y. Escapement $\bar{P}_O$ is the one that provides the maximum sustainable yield Y_O

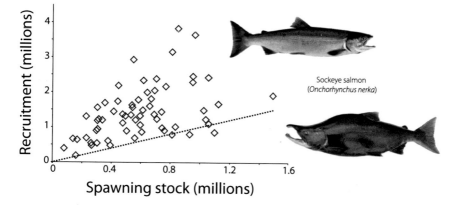

Fig. 8.18 Spawning stocks and corresponding recruitments for the sockeye salmon of the Skeena River (British Columbia, Canada) between 1908 and 1970 (after Walters 1975). The dotted straight line is the 45° line (note the different scales of the two axes). During different stages of life, the sockey salmon can show silvery colors (ocean phase, female on the top picture), which change dramatically towards reddish after entering rivers for spawning (freshwater phase, male on the bottom)

of the Skeena River (British Columbia, Canada) from 1908 to 1970. In theory, these data should be well described by a Ricker curve, while actually such a curve may perhaps provide an idea of the median recruitment corresponding to each spawning stock.

More realistic analyses of the relative merits of harvesting policies would therefore require the explicit introduction of this variability into the demographic model of the population under study. This can be done by introducing a probabilistic approach to the problem of optimal management, e.g. via stochastic models such as those discussed in Chap. 3. For instance, to each spawning stock more than one recruitment can correspond, each with different probabilities. Using a model of this kind, Gatto and Rinaldi (1976) showed that the optimal constant escapement policy ensures a slightly higher average yield than the optimal constant effort policy, but that in general this has the drawback of producing a higher variance of the yield, which may be not socially acceptable as we pointed out above.

8.5   The Bio-Economic Viewpoint

The approach to the rational management of animal or plant harvesting as illustrated in the previous section makes use of purely biological concepts that cannot satisfy economists for a number of reasons.

The first reason is that measuring the productivity of activities that are undoubtedly economic, like hunting, fishing or forestry, in terms of sheer biomass yield is totally insufficient, rather they should be evaluated for the economic and social benefits they provide. In monetary terms this requires the introduction of suitable prices and costs.

A second reason for dissatisfaction lies in the fact that the point of view illustrated in the previous section assumes that the renewable resource to be managed is the property of a private owner or of a single institution, or alternatively that there exists an authority which is strong enough as to impose its decisions to the exploiters. However, as we already pointed out at the beginning of this chapter, many biological resources are often open-access owing to their intrinsic features. A centralized control, even if commendable in principle, is not always possible. For instance, as we said above, certain fish stocks are not in the territorial waters of any country. Even if exclusive economic zones (EEZ) have been created that extend 200 miles from the outer limit of territorial sea, where resource exploitation is restricted to the country controlling the EEZ, some fish stocks are distributed over different EEZ's or outside any EEZ. Economists also may question the social and economic efficiency of centralized planning for the exploitation of renewable resources. Does this kind of planning maximally contribute to the welfare of a country?

The first detailed analysis of the economic problems associated with the exploitation of renewable resources is the one made by Gordon (1954). He was requested by the Canadian government to find out why that country's fisheries were both overexploited and economically inefficient. We will briefly illustrate his theory. To describe the dynamics of the exploited population, we can use the model (8.1), namely

$$\dot{N} = NR(N) - qEN.$$

Suppose that each biomass unit that is harvested is sold at a price p constant over time and independent of the quantity of available resource. These assumptions, although they considerable simplify a real market economy, are reasonable, at least as a first approximation. According to this hypothesis, the exploiters (considered in their totality) would have a continuous flow of revenues, which at time t would be given by $pqE(t)N(t)$. If then we assumed that each effort unit (for example, the work day of a lumberjack or of a vessel crew) has a constant cost c, there would be a flow of costs $cE(t)$. The total profit or net economic return of the harvesting activity would be thus given by $pqE(t)N(t) - cE(t)$.

Gordon conducted a quasi-equilibrium analysis, namely he assumed that effort varies so slowly that the system can be considered at the stable equilibrium for any given level of effort. In other words, one can consider the sustainable profit resulting from the revenues that are obtained by assuming that the model of resource dynamics is at equilibrium. Assume that $E(t) = \bar{E} =$ constant, then the total sustainable profit Π is given by

$$\Pi = pY - c\bar{E} = pq\bar{E}\bar{N} - c\bar{E}. \tag{8.16}$$

From a graphical point of view, the profit can be simply obtained from the effort-yield diagram such as that shown in Fig. 8.19. Because we assumed a constant price, the curve of revenue flows has the same unimodal shape as the curve of yields, while the costs increase linearly with effort.

It is particularly significant to highlight the effort $\bar{E}_B$ at which the revenue curve intersects the line of costs. In fact, for effort values higher than $\bar{E}_B$, the profit is negative because the sustainable harvest is small and the ensuing revenues are thus insufficient to cover the high cost of effort. The opposite obtains if the effort is less than $\bar{E}_B$. The effort $\bar{E}_B$ and the corresponding population size $\bar{N}_B$ are easily calculated by imposing that Π vanishes, i.e.

$$\Pi = 0 = \left(pq\bar{N}_B - c\right)\bar{E}_B$$

whence

$$\bar{N}_B = \frac{c}{pq}. \tag{8.17}$$

Moreover, since

$$\bar{N}_B R(\bar{N}_B) = q\bar{E}_B\bar{N}_B$$

Fig. 8.19 Revenue curve
(pY) and cost line $(c\bar{E})$ as a
function of effort. The value
$\bar{E}_B$ is the effort at bionomic
equilibrium (see text)

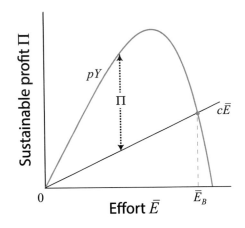

we obtain

$$\bar{E}_B = \frac{1}{q} R\left(\frac{c}{pq}\right). \tag{8.18}$$

Suppose there is open access to the resource, namely exploitation is unregulated. A legitimate question is whether there is a value of the effort that will establish in the long term and what this effort may be. According to Gordon (1954), as there is no restriction to the entry of new exploiters, if the total effort is less than $\bar{E}_B$, since the total profit Π is positive, as well as the profit per unit effort, other lumberjacks or hunters or fishermen have an incentive to access to the resource exploitation. Therefore, effort tends in such a case to increase. If, on the contrary, effort is larger than $\bar{E}_B$, the total profit Π is negative; so, some of the exploiters must be at a loss and have an incentive to abandon the activity, thus decreasing the total effort devoted to harvesting the resource. In conclusion, Gordon predicts that there exists an effort value which tends to settle in the long term and this value is $\bar{E}_B$, in correspondence of which the total profit from the economic activity vanishes. This long-term situation is termed *bionomic equilibrium* in the economic literature.

That the profit is zero in the long run if there is no regulation can appear to be an extravagant outcome: in fact, on what would these exploiters make a living if their revenues equal the costs? However, this long-term situation can be better understood if we specify that the cost c is correctly defined as an *opportunity cost*. With this term economists refer to the actual cost of an effort unit (for example, the cost of a workday by a fisherman) increased by the remuneration that such an effort unit would get in the most profitable alternative. In the case of the above fisherman, one would add to the cost of the workday—dictated by labour availability and market conditions—the cost of not devoting that same workday to another alternative activity which would provide a certain remuneration. It follows that the profit at bionomic equilibrium is not zero in real terms, but is the minimal remuneration that would not induce effort units to abandon the activity of resource exploitation and devote labour and capital to other activities. If one considers that renewable resource exploitation is often exerted in many developing countries, one can understand that the alternative economic activities may be very scarce and labour costs very low, thus implying that the profit which establishes under open access to the resource and the corresponding population size $\bar{N}_B = \frac{c}{pq}$ can indeed be quite close to zero in the long run.

Thus the analysis suggested by H.S. Gordon agrees with the fishermen's dilemma proposed by Colin Clark (1981): in the absence of some form of regulation the sustained economic performance resulting from the exploitation of a renewable resource is very low and there is economic inefficiency. In fact, the resource harvesting does not add to a country richness because the same monetary profit could be generated by another economic activity. In addition, there is often biological overexploitation. In particular, it must be remarked that the bionomic population size $\bar{N}_B = \frac{c}{pq}$ is not at all related to the demography of the population being harvested, but only to price, cost and catchability coefficient. The higher the ratio between prices and costs, the smaller the population size and the higher the extinction risk. Undoubtedly, Gordon's theory is a simplification with respect to reality, but it explains the basic phenomena that have led many renewable resources to both biological and economic disaster. After a tumultuous initial period of increasing effort, characterized by profitability for all economic operators, a period follows during which harvested biomass decreases, and thus ensuing profits, exploitation activities are downsized and finally settle to levels that allow for just modest profitability. In some cases, the downsizing occurs too late and the result is a real biological and economic catastrophe. Suffice it to remember the collapse of whaling (see Fig. 8.20 drawn after Horwood 1981) and

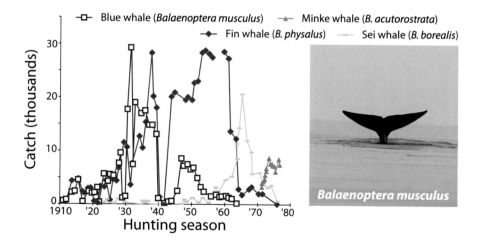

Fig. 8.20 Catches of the four main species of baleen whales hunted in the past century and belonging to the genus *Balaenoptera*: blue whale (*Balaenoptera musculus*, blue squares), fin whale (*Balaenoptera physalus*, red diamonds), sei whale (*Balaenoptera borealis*, yellow dashes) and Minke whale (*Balaenoptera acutorostrata*, green triangle). The inset shows a blue whale diving in the Monterey Bay National Sancturay of the Monterey bay (California, US). Data are after Horwood (1981)

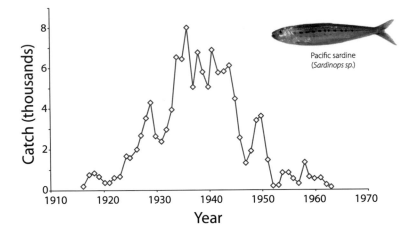

Fig. 8.21 Yearly sardine (*Sardinops caerulea*) catches along the Pacific shores of North America. Data after Murphy (1966)

the collapse of the above cited fishery of the Peruvian anchoveta in 1973 and of the California sardine in the 1960s (Murphy 1966), as shown in Fig. 8.21. Suffice it to mention, as regards Italy, the collapse or decline of several Adriatic fish stocks, not to mention all the problems caused by indiscriminate hunting supported by vanishing effort costs, which has led to the practical extinction of several populations of birds and mammals. Figure 8.22 shows the data regarding the Adriatic stock of the European hake (*Merluccius merluccius*, Levi and Giannetti 1973). Note that the figure reports the catch per unit effort. In fact, if we assume the validity of the relationship

$$h = qEN$$

whence

$$\frac{h}{E} = qN$$

it turns out that catches per unit effort are proportional to the abundance of the population (which is quite intuitive) and are therefore a good indicator of a stock "health". As one can notice the hake catches per unit effort have been steadily decreasing between 1958 and 1971 with a reduction from 3 to 1.

It is worthwhile to note that the effort that establishes at bionomic equilibrium does not necessarily lead to overexploitation of biological resources. In principle, it might not be true that $\bar{E}_B$ is greater than the effort $\bar{E}_0$ that ensures the maximum sustainable yield of biomass. If the costs of effort are sufficiently high, in fact, a situation may occur like the one illustrated in Fig. 8.23a. It must be emphasized, however, that this case is quite rare, because clearly there is little interest to exploit a resource that involve high costs of effort with respect to selling prices.

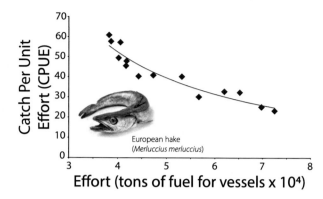

Fig. 8.22 Efforts and catches per unit effort for the fishery of European hake in the Adriatic between 1958 and 1971 (Levi and Giannetti 1973). Effort is measured as annual fuel consumption of the fishing vessels

At this point the question arises of whether there exist tools that can be used and policies that can be put in place to avoid the inefficiency of the bionomic equilibrium. Effort could indeed be theoretically calibrated so as to obtain the *maximum sustainable profit*, i.e. so as to maximize Π. Therefore, one should find the value $\bar{E}_M$ of effort at which

$$\max \Pi = p\bar{N} R\left(\bar{N}\right) - c\bar{E}.$$

The optimal effort $\bar{E}_M$ (see Fig. 8.23) is lower than both effort $\bar{E}_0$ that guarantees MSY and effort $\bar{E}_B$ at bionomic equilibrium. From a mathematical viewpoint, $\bar{E}_M$ is the value of effort at which the tangent to the curve of revenues pY is parallel to

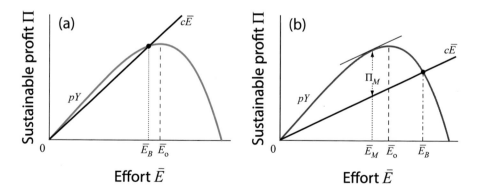

Fig. 8.23 **a** If the opportunity cost of one unit of effort is high enough, the effort $\bar{E}_B$ at bionomic equilibrium can be lower than the effort $\bar{E}_0$ that guarantees the maximum yield of biomass. **b** For given revenue and cost curves as a function of effort, $\bar{E}_M$ is the effort that provides the maximum sustainable profit, namely the one at which $\max(\Pi)$. With $\bar{E}_0$ we indicate the effort that maximizes the biomass yield—namely, the one at which $\max(Y) = \max(NR(N))$. $\bar{E}_B$ is the effort at bionomic equilibrium ($\Pi = 0$)

the line of costs $c\bar{E}$, or, as economists say, is the value of effort at which marginal revenue equals marginal cost. The corresponding maximum profit is indicated by Π_M in Fig. 8.23b. Note that sustainable profit maximization can lead to lower resource exploitation than the nullification of the same profit or the maximization of biomass yield (the MSY concept that we previously discussed). It is to be remarked, however, that this result holds true only if we neglect an important economic parameter, namely the discount rate, which is a way of providing less value to future uncertain revenues. This concept will be clarified in the next section.

If we adopt the simple Schaefer model (8.12) with logistic growth

$$N R(N) = rN \left(1 - \frac{N}{K} \right)$$

we can explicitly calculate both $\bar{E}_M$ and $\bar{N}_M$. In this case, the effort at bionomic equilibrium is given by

$$\bar{E}_B = \frac{r}{q} \left(1 - \frac{c}{pqK} \right).$$

Recalling that the sustainable yield is provided by (8.14), i.e.

$$Y = qK\bar{E} \left(1 - q\frac{\bar{E}}{r} \right)$$

one obtains that the sustainable profit as a function of effort $\bar{E}$ is given by

$$\Pi = pqK\bar{E} \left(1 - q\frac{\bar{E}}{r} \right) - c\bar{E} \tag{8.19}$$

Therefore the effort $\bar{E}_M$ providing the maximum sustainable profit is the solution of the equation

$$\frac{d\Pi}{d\bar{E}} = pqK \left(1 - 2q\frac{\bar{E}}{r} \right) - c = 0$$

namely

$$\bar{E}_M = \frac{r}{2q} \left(1 - \frac{c}{pqK} \right). \tag{8.20}$$

Considering again the case of the yellowfin tuna, for example, Schaefer (1967) assesses the stock size $\bar{N}_B$ at bionomic equilibrium to be equal to $0.5K$ (see Fig. 8.14). As $\bar{N}_B = \frac{c}{pq}$, it follows that the ratio $\frac{c}{pqK}$ is equal to $\frac{1}{2}$ and thus

$$\bar{E}_B = \bar{E}_0 = \frac{r}{2q} = 34,300 \text{ fishing days per year}$$

$$\bar{E}_M = \frac{r}{2q}\left(1 - \tfrac{1}{2}\right) = 17,150 \text{ fishing days per year.}$$

There remains the problem of how to achieve in practice the goal of a sustained effort that ensures economic efficiency. One possibility is that of the direct control of effort exerted by an appropriate institution. Such a result could for example be achieved by introducing limitations to the fishing or hunting season. The drawback of a similar measure (highlighted by Table 8.1 with reference to the case of the Peruvian anchoveta, *Engraulis ringens*) is that it does not prevent the increase of the number of exploiters, thus leading to the unpleasant result of a very limited season with a large number of fishermen or professional hunters being idle for most of the year. Also, if the yield must be transformed (e.g. canned) via suitable industrial plants, these would be functioning only during the peak period. Even the alternative of granting hunting or fishing licenses, in order to limit the number of exploiters, is not without drawbacks. First of all, there is a problem of fairness: the resource exploiter who is given a license by the government receives a privilege that implies an economic benefit that is denied to others. Auctioning licenses is a possibility, but so far no government has dared to do so with the exception of sport fishing or hunting. A second problem is that in any case those who have been granted licenses are competitive and thus unable to escape the fishermen's (or hunters' or lumberjacks') dilemma. They have anyway a tendency to increase their individual effort. One third and final problem which has been already mentioned in Sect. 8.2 is that the increase of the catchability coefficient (due to technological progress) can frustrate the effort limitation operated through licenses.

As an alternative to the direct control of effort, a regulatory method that is favourably looked upon by many economists is to levy taxes on the amount of harvested resource. If one denotes by τ the tax levied on each unit of harvest, the profit becomes

$$\Pi = (p - \tau)Y - c\bar{E}.$$

Therefore, if this is the only form of regulation, it is likely that a bionomic equilibrium will be reached for which the usual formula applies provided one replaces price p with $p - \tau$. By suitably selecting the tax it is possible, for example, to bring the effort at the taxed bionomic equilibrium down to the effort that maximizes the sustainable profit in absence of any tax. This outcome is graphically illustrated in panel (a) of Fig. 8.24. Despite its rationality, the tax system is rarely considered seriously, with the exception of some cases of big game hunting, in which hunters must pay a fee for each slaughtered animal. There is indeed a strange tendency to consider with suspicion those tools of environmental regulation that are of a fiscal rather than coercive nature. Suffice it recall that levying taxes on fishing has even been forbidden by law in the United States since 1976.

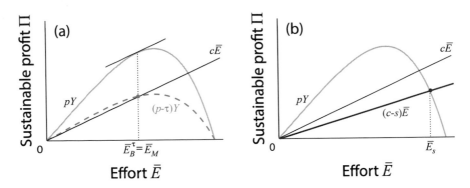

Fig. 8.24 Effects of levying taxes or providing subsidy on harvesting regulation. **a** By suitably choosing the tax τ, it is possible to let the effort at the taxed bionomic equilibrium coincide with the effort that would maximize the sustainable profit without levying any tax. **b** The effect of a subsidy s provided by government to each effort unit. The features of the new bionomic equilibrium that establishes in the long run are discussed in the main text

A drawback of taxing the harvested biomass is that with this system the exploiters' net return is anyway dissipated. There is a net benefit to the society, but that is represented by the tax income which equals τY, where Y is the sustainable yield. However, it would be possible to indirectly return this income into the pockets of hunters or fishermen, for example by investing the levied tax into improving their living conditions (construction of roads, schools, etc.). A second serious inconvenience is that the tax system cannot be used for resources under the jurisdiction of more than one country, for diplomatic reasons. Even when there exist international commissions of control for the exploitation of resources that are shared by many countries, they have never been given the power to collect taxes.

Far from imposing taxes, many governments are prone to giving subsidies, which are in many cases proportional to the effort. For instance, the Italian government provided fuel to fishing vessels (diesel fuel) at a reduced price. The rationale behind the subsidy is that these economic activities should be supported, because the profit is minimal. And indeed it is, because we know that the net economic benefit is dissipated at bionomic equilibrium, but the cause is too much effort, not too little. What is the result of this policy of subsidizing effort? Let s be the subsidy per unit effort; then the profit is given by

$$\Pi = pY - c\bar{E} + s\bar{E}$$

and, as is clear from panel (b) of Fig. 8.24, the long-term result is that the system converges to a new bionomic equilibrium with a corresponding effort that is even higher, a resource even more exploited and a profit equally null. In addition the taxpayers must bear the cost of the subsidy. The short-sighted policy of subsidizing the harvesting of renewable resources, in particular fishing, is never deprecated enough. It led to the impoverishment of our biological fish stocks and the economic inefficiency of fishing activities.

In conclusion, the exploitation of biological resources needs some control by an independent authority, otherwise economic inefficiency will certainly occur. However, this control is hard to be effectively implemented and there is still much debate about what the best policy should be in order to limit the exploitation effort so as to ensure economic profitability and suitable protection of the animal or plant population subject to exploitation. The economic literature on these topics has been developed in the past decades and the reader interested to deepening these aspects should have a look at it. A fundamental book, even if it was written four decades ago, is still the one by Clark (1976).

8.6 Harvesting Populations with Age and Size Structure

The approach followed so far does not fit into those cases where there is a remarkable inhomogeneity between the individuals of an exploited population. This is indeed the case in which the biomass, thus the commercial value of each individual (tree, fish or mammal), varies significantly with age. Obviously, the size increases with age, and so does the individual biomass, but the value per unit biomass does depend on age in many cases. So, for instance, trees below a certain age, consequently below a certain size, cannot provide usable timber, while on the other hand the larger the tree, the higher the value of the harvested timber. Therefore, total numbers or total biomass can no longer be used as the only variables that can aptly describe the state of a population.

A comprehensive analysis that accounts for both age classes and density dependence is a bit complicated and beyond the scopes of this text (the interested reader may consult the paper by Reed 1980). To facilitate the analysis, we consider here a case that is in a sense dual to the one analysed in previous sections, namely the one in which dependence on density can be neglected as regards both survival and reproduction, while the differentiation into different age classes plays a fundamental role. In particular we will treat two simple, but important cases: (a) that of a managed forest (or aquaculture plant) for which the optimal age must be chosen at which the trees (fish) should be harvested, (b) that of a natural (or semi-natural) fish population with constant recruitment for which the effort and the selectivity of the fishing device must be decided upon.

8.6.1   Optimal Rotation Period of a Renewable Resource and the Problem of Discounting

The problem is that of planning the regular tree cut in a forest where the density is not so high as to result in mortality due to intraspecific competition. The same paradigm applies to an aquaculture plant where food is regularly provided to e.g. fish or crustaceans which, after being stocked at young age, then grow in age and body size. From now on, we will refer to forestry and trees, but it should be clear that the same basic concepts can be applied to aquafarming too. Problems of the next chapter further clarify this aspect.

If the forest is subject to silvicultural practices, reproduction does not occur naturally but is replaced by planting of new trees after cutting down a forest stand. Additionally, it is justified to assume centralized management, because in most cases woods are privately owned or are state property, but are given to privates as a concession, or are directly managed by public authorities.

The basic fact is that the biomass of each tree, and thus the obtained timber production, varies with age. A prototype yield curve as a function of age is shown in Fig. 8.25 for *Pinus radiata*, the Monterey pine. The curve is obviously growing but saturates as it reaches a *plateau*. At old ages there can also be a decreased timber production. Parallel to production, one can can consider, as a function of age, the commercial value of a tree or better of an even-aged stand, that is an area with trees that are all the same age, namely that are a cohort. Curves of revenues and costs per acre in the already mentioned case of *Pinus radiata* stands are also shown in Fig. 8.25. The revenue curve is in this case proportional to that of the total timber production, which implies that, once the age of usable timber is reached, the price per unit timber volume of this pine is constant. As for the curve of silvicultural costs, of course they increase with the stand age. In addition to initial fixed cost for the planting of new young trees, in fact, there are maintenance costs that accumulate

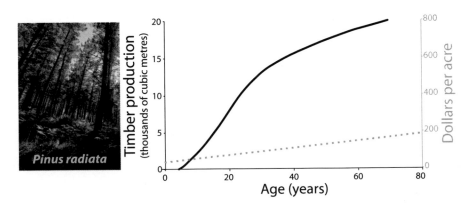

Fig. 8.25 Total timber production (solid curve and left ordinate axis), revenues (same curve, but right axis) and costs (dotted straight line, right axis) as a function of stand age for an acre of Monterey pine *Pinus radiata* (Gregory 1972)

year after year. Finally, there is a cost for the final tree felling which usually can be assumed to be fixed, not dependent on tree age.

The simplest harvesting problem in forest exploitation is to choose the optimal rotation period, namely to decide how often an even-aged stand should be cut down (in aquaculture, how often a fishpond should be emptied). If we denote by $V(t)$ the commercial value of a stand of trees all with the same age t, by $c(t)$ the accumulated planting and maintenance costs, and by c_A the cost of clear-cutting the stand (which we assume to be independent of age), then the profit that one gets at the end of a period of duration T is

$$\Pi_{tot}(T) = V(T) - c(T) - c_A.$$

Note that this profit is obtained after a number of years T during which no profit from the cohort of trees is actually gained. Therefore, to determine the rotation period, one should not consider the profit Π_{tot} obtained in that year, rather the average profit Π_{ave}, that is

$$\Pi_{ave} = \frac{\Pi_{tot}}{T} = \frac{V(T) - c(T) - c_A}{T} \tag{8.21}$$

The optimal rotation period T_o is found by searching for the maximum average profit, which can be done by equating its derivative to zero. From relationship (8.21), one gets

$$\frac{\left(V'(T) - c'(T)\right)T - \left(V(T) - c(T) - c_A\right)}{T^2} = 0$$

or, equivalently,

$$\frac{V(T) - c(T) - c_A}{T} = \frac{d}{dT}\left(V(T) - c(T) - c_A\right) \tag{8.22}$$

Therefore, the optimal rotation period is the age at which the average profit equals the marginal profit.

Figure 8.26 shows a simple graphical interpretation of condition (8.22). Table 8.4, for example, shows that the optimal rotation period for Douglas fir in British

Fig. 8.26 Graphical determination of the optimal rotation period for a forest stand: marginal profit equates the average one. If the interest rate is zero, the optimal rotation period is T_o, whereas, if the interest rate is infinite, the optimal period is T_∞

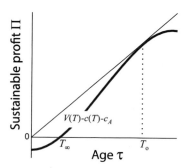

Table 8.4 Profits that can be obtained from a typical stand of Douglas fir in British Columbia, Canada (Pearse 1967)

Age (years) T	Net stumpage value ($) $\Pi_{tot} = V(T) - c(T) - c_A$	Average profit ($/year) $\Pi_{ave} = \frac{\Pi_{tot}}{T}$
30	0	0
40	43	1.08
50	143	2.86
60	303	5.05
70	497	7.10
80	650	8.12
90	805	8.94
100	913	9.13
110	1,000	9.09
120	1,075	8.93

Columbia is 100 years. Of course, one should not think that profits are obtained only once every 100 years. A forest is a mosaic of several cohorts, namely several stands each consisting of trees with different ages; 100 years is therefore the optimal period for each cohort.

The extreme simplicity of dealing with this problem allows us to introduce an economic aspect of the utmost importance that we have hitherto neglected. In all the cases in which there is a long delay between the moment when capital is invested (tree planting) and the time at which the earnings from that capital (tree cut) are appropriated, it is not fair, from an economic viewpoint, to assume that sums of money spent or earned today are completely equivalent to sums spent or earned in a more or less distant future. This problem should not be confused with that of inflation, which is a depreciation of any currency due to a general increase of market prices. All the sustainable revenues and costs we are talking about are actually deflated, that is they are purged of inflation rates. What we mean is that profits expected for the future are uncertain and this is much more so if the future is distant in time. Moreover, those who invest capital in a productive activity might decide for an alternative investment and thus expect that their capital increases over time with an interest rate at least equal to the prevailing earnings from money investment (this is also termed the opportunity cost of capital).

That is why the value of future earnings, from a purely monetary perspective, must be discounted so as to obtain the so-called present value. More precisely, if Π_k is a profit that will be obtained k years from now and i is the annual interest rate, the present value of this profit is given by

$$V_\Pi = \frac{\Pi_k}{(1+i)^k}.$$

Indeed, if V_Π had been invested at compound interest, after k years one could collect exactly the sum Π_k. It is also useful to introduce the instantaneous rate of discount δ, defined by the equation

$$\exp(\delta) = 1 + i$$

i.e.

$$\delta = \log(1 + i).$$

By using the instantaneous discount rate, the present value of a profit Π obtained after T years is

$$V_\Pi = \Pi \exp(-\delta T).$$

We now introduce these concepts into the problem of determining the optimal rotation period. For simplicity, we assume that the costs of planting and maintenance are negligible, even if the analytical treatment could be easily extended to the general case. So, the profit that can be gained every T years is worth $V(T) - c_A$. Hence, the present value of all profits that can be periodically obtained since year 0 when the forest was first planted is given by

$$V_\Pi = \sum_{k=1}^{\infty} \exp(-k\delta T)(V(T) - c_A).$$

Using the formula for summing geometric series we obtain

$$V_\Pi = \frac{V(T) - c_A}{\exp(\delta T) - 1}.$$

Then, the optimal rotation period is obtained by searching for that value of T that maximizes the present value. It is the solution of the following equation

$$\frac{d(V_\Pi)}{dT} = \frac{V'(T)\left[\exp(\delta T) - 1\right] - \delta \exp(\delta T)(V(T) - c_A)}{\left[\exp(\delta T) - 1\right]^2} = 0$$

namely

$$\frac{V'(T)}{V(T) - c_A} = \frac{\delta}{1 - \exp(-\delta T)}. \tag{8.23}$$

Equation (8.23) is termed *Faustmann's formula* from the name of the German forester who first derived it Faustmann (1849). As shown by Table 8.5, which still refers to Douglas fir, the effect of the discount rate is to significantly decrease the rotation period. In the limit of $\delta \to \infty$ the optimal period T_o is given by

$$V(T_\infty) - c_A = 0$$

Table 8.5 Optimal rotation period and annual average profit as a function of various discount rates for the Douglas fir in British Columbia, Canada

Annual interest rate i (%)	Optimal rotation period (years)	Annual average profit ($/year)
0	100	9.1
3	70	7.1
5	63	5.6
7	56	4.2
10	49	2.8
15	43	1.7
20	40	1.2

that is to say that a tree is cut as soon as it reaches a commercial value that equals the opportunity cost c_A (see again Fig. 8.26). Obviously, for intermediate values of δ, the optimal rotation period is in between T_0 (corresponding to $\delta = 0$) and T_∞.

Note that the period T_∞ is the one that would also be established in the long run in the case of an open-access forest, which is not privately owned or centrally managed but left open to exploitation without any regulation. This is not infrequent, specially in developing countries. The reasoning that justifies this outcome is exactly the same as the one we reported earlier when we illustrated H. S. Gordon's theory. If profits are positive, there is indeed a tendency to increase the harvesting of younger trees by the exploiters who last access the resource. Conversely, if profits are negative.

From this simple example we observe that the open-access situation coincides, in the limit, with the situation of centralized management that cares little or nothing for the future (discount rate $\delta \to \infty$). Although illustrated for the simple case of optimal rotation in a forest, this conclusion is quite general and also applies to other problems of renewable resources management.

8.6.2 Managing the Harvest of Fish Populations with Constant Recruitment

In many fish stocks the number of individuals of the first catchable age class (the so-called recruitment) is relatively constant over time. This can be due mainly to two causes. The first is that sometimes the recruitment is practically independent of the parental stock; this phenomenon was pointed out when we introduced the Beverton-Holt stock-recruitment curve: in very fertile demersal species, such as plaice, the intraspecific competition among larvae and juveniles is so strong as to create a bottleneck. Unless the population size is very small, there exists practically no correlation between the recruitment and the parental stock. The second reason is that in some fish stocks reproduction occurs in areas totally different from those where fishing takes place, and thus the recruitment to a fishing ground is substantially due to the migration of a small fraction of juveniles to the area where the fishing effort is exerted. Unless the fish species is simultaneously overexploited in all the fishing grounds, its reproduction rate remains largely unchanged as well as the fraction that migrates to the fishing ground (i.e. the recruitment). This was the case of European eels, for example. The whole species reproduces in the Sargasso sea, then the larvae (the so-called leptocephali) are transported by the Gulf Stream to the European shores where they are recruited to several lagoons or freshwater water bodies as glass eels (after a metamorphosis). Glass eels then grow and become elvers and yellow eels; when the eels mature sexually, they metamorphose to silver eels and go back to the Sargasso Sea, thus closing the life cycle. Yellow and silver eels are fished in the water bodies where they spend the continental phase of their life. The recruitment to each fishery is thus a tiny portion of the whole population. As a first approximation, one can think of the Sargasso Sea as a very big reservoir that provides more or less the same recruitment to each fresh or brackish water body independently of the fishing effort in a single fishery, unless the effort is simultaneously exerted in very many body waters, thus leading to global overfishing. Unfortunately, in recent years (since the 1970s) there has been a marked decline of anguillid species possibly due to changes in oceanic circulation, parasite infections, habitat disruption, chemical contamination and overharvesting.

If recruitment (that is the size of the first age class on which effort is exerted) is constant, the two most important phenomena are (a) the increase in weight, and therefore in value, of each individual fish for increasing age, and (b) the parallel action of mortality which, as the fish age, decreases the number of individuals available for capture. These two phenomena act in opposite directions: one increases, the other decreases the total fishable biomass (and the total economic value of the potential catch). To quantitatively approach the problem, first consider the situation in which there is no exploitation. In this case, the dynamics of the population can easily be obtained.

Let $n(t, \tau)$ be the number of individuals of age τ at time t and $\mu(\tau)$ the mortality rate at age τ. Because the recruitment is constant, each fish cohort is the same and thus

$$n(t, 0) = R$$

where with 0 we denote the recruitment age and with R the constant recruitment. The time evolution of $n(t, \tau)$ is very simple. Indicating with $p(\tau)$ the probability that a recruit survives from age 0 up to age τ, namely

$$p(\tau) = \exp\left(-\int_0^\tau \mu(\zeta)\,d\zeta\right),$$

we obviously get

$$n(t, \tau) = n(t - \tau, 0)p(\tau) = Rp(\tau). \tag{8.24}$$

In fact, the number of individuals in each age class is independent of time t: because recruitment is constant, the population age distribution is stationary and given by the survival curve of a cohort. Let $w(\tau)$ be the average weight of a fish of age τ. Then at any time t the biomass in each age class, when no fishing takes place, is given by

$$B_N(\tau) = Rp(\tau)w(\tau). \tag{8.25}$$

Because $p(\tau)$ is a decreasing function of age and $w(\tau)$ a growing function, usually $B_N(\tau)$ is unimodal, first increasing and then decreasing with τ. A typical curve for the stock of North Sea plaice (*Pleuronectess platessa*) is shown in Fig. 8.27. Very important is the age τ_0 corresponding to the largest biomass (about 14 years in the case of plaice); it is obtained by setting the derivative of $B_N(\tau)$ to zero, or

$$\frac{dp}{d\tau}w + p\frac{dw}{d\tau} = 0.$$

Since $\frac{dp}{d\tau} = -\mu p$, the equation reduces to

$$\frac{1}{w}\frac{dw}{d\tau} = \mu. \tag{8.26}$$

We now investigate the consequence of introducing exploitation of the fish stock. The fishing gears are usually characterized by specific selectivity, i.e. they differentially act on the various fish sizes, and consequently ages. A brute-force, but fairly realistic, assumption is that selectivity is knife-edge, in other words that, in dependence of the mesh size of the fishing net, there is an age τ_s below which the smaller fish escape capture and above which all the fish that encounter the net are harvested. This is equivalent to assuming that for $\tau < \tau_s$ only the natural mortality $\mu(\tau)$ is operating, while for $\tau \geq \tau_s$ fishing mortality is added on the top of natural mortality.

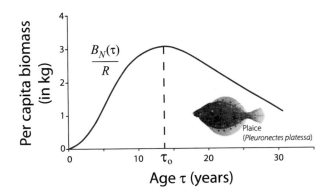

Fig. 8.27 The natural biomass curve at different ages for the North Sea plaice stock (Clark 1976). τ_0 is the age at which biomass is maximum

As usual, the fishing mortality rate is supposed to be proportional to the fishing effort (e.g. measured as number of operating nets) through a catchability coefficient; so for $\tau \geq \tau_s$ the total death rate per capita is $\mu(\tau) + qE$.

In this way we can calculate the biomass production that is obtained by exerting a constant effort via fishing gears that select a certain age τ_s. Following a way of reasoning similar to those previously used, we easily get the number of individuals of age τ at time t. It is given by

$$n(t, \tau) = \begin{cases} Rp(\tau) & \text{for } 0 < \tau < \tau_s \\ Rp(\tau_s)p(\tau - \tau_s)\exp\left(-q\bar{E}(\tau - \tau_s)\right) = Rp(\tau)\exp\left(-q\bar{E}(\tau - \tau_s)\right) & \text{for } \tau \geq \tau_s \end{cases}$$

Consequently, the standing biomass curve for the population subjected to fishing is

$$B(\tau) = \begin{cases} B_N(\tau) & \text{for } \tau < \tau_s \\ B_N(\tau)\exp\left(-q\bar{E}(\tau - \tau_s)\right) & \text{for } \tau \geq \tau_s \end{cases}$$

and is thus easily obtained from the natural biomass curve (see Fig. 8.28). The total biomass yield (per unit time) is then obtained as the sum of all contributions of biomass flows harvested from every age class with $\tau \geq \tau_s$. More precisely, the sustained yield Y is provided by the formula

$$Y = \int_{\tau_s}^{+\infty} q\bar{E}B(\tau)d\tau = q\bar{E}\exp\left(q\bar{E}\tau_s\right)\int_{\tau_s}^{+\infty} B_N(\tau)\exp\left(-q\bar{E}\tau\right)d\tau. \quad (8.27)$$

This formula expresses, for a given selected age τ_s, the biomass yield as a function of effort: $Y = Y_s(\bar{E})$. If $\bar{E}$ is zero the yield obviously vanishes; instead, if $\bar{E} \to \infty$ the

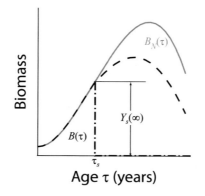

Fig. 8.28 Biomass of a constant-recruitment hypothetical stock at different ages under natural conditions (solid blue curve) and under constant-effort exploitation. The dashed curve corresponds to a finite fishing effort, while the dash-and-dot line corresponds to an infinite effort. In both cases the adopted mesh size of the fishing gears selects the minimum age τ_s

yield does not vanish, as one might expect, but is equal to $B_N(\tau_s)$, as is intuitively clear from the dash-and-dot curve of Fig. 8.28. All biomass is harvested at the minimum age τ_s; as soon as a younger age class reaches τ_s, its entire biomass is harvested. Of course, this result should be taken with caution. Exerting a very large effort, although it is restrained to age classes older than the selected age, might imply an overexploitation of the fish stock, thus resulting in a violation of the assumption that the recruitment (given e.g. by a Beverton-Holt stock-recruitment function) is constant.

Depending on the value of τ_s, $Y_s(\bar{E})$ may be an increasing or unimodal function of $\bar{E}$, as exemplified in Fig. 8.29 which still makes reference to the stock of North Sea plaice.

In a similar way, given the effort $\bar{E}$, there exists an optimal value of the gear mesh size (i.e. a value of the selected age τ_s), which is the one that ensures the maximum sustainable yield MSY. Beverton and Holt (1957), who pioneered the theory illustrated above, have introduced the concept of eumetric yield curve (from the Greek term *eu-metros*, namely well-proportioned). It is obtained by recording, for each effort $\bar{E}$, the largest sustainable yield obtainable with the corresponding optimal mesh size. The eumetric curve is clearly the envelope of the individual yield curves

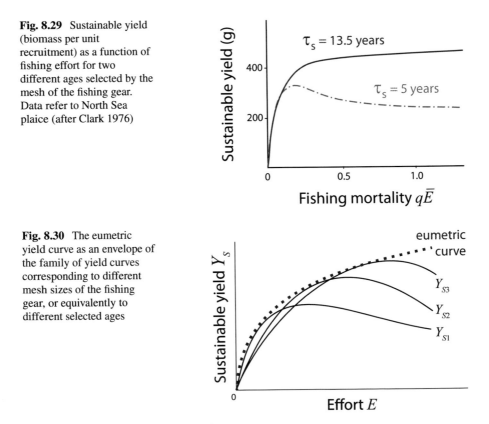

Fig. 8.29 Sustainable yield (biomass per unit recruitment) as a function of fishing effort for two different ages selected by the mesh of the fishing gear. Data refer to North Sea plaice (after Clark 1976)

Fig. 8.30 The eumetric yield curve as an envelope of the family of yield curves corresponding to different mesh sizes of the fishing gear, or equivalently to different selected ages

$Y_s(\bar{E})$ as illustrated in Fig. 8.30. Note that the eumetric yield is an ever increasing function of effort. In fact the situation that, theoretically, provides the maximum sustainable yield is the one in which an infinite effort is exerted with a mesh size selecting the age $\tau_s = \tau_0$ (namely the age at which the natural standing biomass is maximum) as made clear in Fig. 8.31. As we stated above, this result should be taken with some caution.

As was explained in the previous section, this analysis in terms of sheer biomass is incomplete from an economic point of view. For instance, exerting an infinite effort implies infinite costs which is not acceptable. It is easy, however, to extend the theory by including costs and revenues. Just let $w(\tau)$ in formula (8.27), which provides the sustainable yield Y, be the average price of a fish of age τ rather than its average weight. Then that same formula provides the total revenue of the fishery. Note that not always the average price is obtained by multiplying weight times a constant price per unit weight. In fact, as is well known, fish of particular sizes (large, medium or small, depending on the fished species) may have different prices per unit of weight. As for the cost of fishing, using different mesh sizes has negligible impact on costs; thus the cost can be simply obtained by multiplying the effort $\bar{E}$ times an opportunity cost c.

Let us first assume that the management is implemented in the best way by a regulatory body (agency, government etc.) that is empowered to impose both a minimum mesh size and a maximal effort. Clearly the revenue as a function of effort can be calculated from the eumetric revenue curve (if the price per unit weight is constant and equal to p, this curve is nothing but that of eumetric yield multiplied by p). The situation is graphically represented in Fig. 8.32, which shows that the effort ensuring the maximum sustainable profit is $\bar{E}_M$. At that effort the eumetric marginal revenue equals the opportunity cost. Of course this is predicated on the optimal eumetric mesh size being imposed by the regulating agency.

What would instead occur if there were no regulation of harvesting, namely if the resource is open-access? Presumably fishermen would tend to use smaller and smaller mesh sizes so as to catch any fish that has commercial value. Thus the revenue curve would correspondingly have the form shown in Fig. 8.33. Also, effort would

Fig. 8.31 Curve of standing biomass at different ages under natural conditions (solid line) and with complete exploitation of all the age classes older than τ_0 (dash). The exerted effort is infinite and the mesh size is chosen in such a way as to select the age at which the natural biomass is maximum

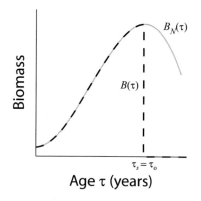

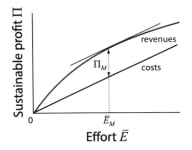

Fig. 8.32 Curves of revenues (under the assumption that yield is eumetric) and costs as a function of sustained effort. Effort $\bar{E}_M$ is the one that guarantees the maximum sustainable profit

increase until profit is completely dissipated in accordance with Gordon's theory (effort $\bar{E}_1$ in Fig. 8.33). In this situation the resource exploitation is negative for two reasons: the mesh size is too small and the effort is too large. Suppose that, as a consequence, a regulatory agency takes over and enforces a eumetric mesh. If the effort remains at $\bar{E}_1$, a positive profit Π is generated in the long run. If, however, the effort is not regulated (in other words the resource is still open-access), the situation would anyway evolve towards a bionomic equilibrium corresponding to the effort $\bar{E}_2$, where the mesh of the nets is eumetric but the effort is excessive and the long-term profit is still zero. The only advantage would be a greater production of biomass.

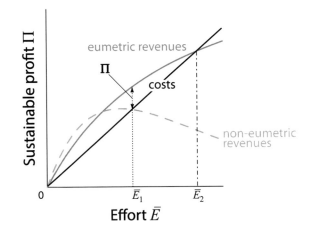

Fig. 8.33 Curves of eumetric (blue solid line) and non-eumetric revenues (grey dashed) as a function of sustained effort. Non-eumetric revenues are due to the use of too small mesh sizes in open-access conditions. Given certain fishing costs, $\bar{E}_1$ is the effort that would obtain at the bionomic equilibrium. If the mesh of the gear is regulated in a eumetric way, but effort is not, the partially regulated bionomic equilibrium is $\bar{E}_2$

References

Byerly M, Brooks B, Simonson H, Savikko B, Geiger HJ (1999) Alaska commercial salmon catches, 1878–1997. Regional Information Report 5J99-05, Alaska Department of Fish and Game, Juneau, Alaska

Clark C (1976) *Mathematical Bioeconomics*. Wiley, New York

Clark C (1981) Bioeconomics of the ocean. *BioScience* 31:231–237

Cooley R (1963) *Politics and conservation: The Decline of the Alaska Salmon*. Harper & Row, New York

Faustmann M (1849) Berechnung des werthes, welchen waldboden sowie nach nicht haubare holzbestande für die weldwirtschaft besitzen. *Allgemeine Forst und Jagd Zeitung* 25:441

Gatto M (1995) Sustainability—is it a well-defined concept? *Ecological Applications* 5:1181–1183

Gatto M, Rinaldi S (1976) Mean value and variability of fish catches in fluctuating environments. *Journal of the Fisheries Research Board of Canada* 33:189–193

Glantz M (1979) Science, politics, and the economics of the Peruvian anchoveta fishery. *Marine Policy* 3:201–210

Gordon H (1954) Economic theory of a common-property resource: The fishery. *Journal of Political Economy* 62:124–142

Gregory G (1972) *Forest Resource Economics*. Ronald, New York

Hilborn R, Quinn TP, Schindler DE, Rogers DE (2003) Biocomplexity and fisheries sustainability. *Proceedings of the National Academy of Science of the United States of America* 100:6564–6568

Horwood J (1981) Management and models of marine multispecies complexes. In: CW F, TD S (eds) *Dynamics of Large Mammal Populations*. Wiley, New York, pp 339–360

Iannelli M, Pugliese A (2014) *An Introduction to Mathematical Population Dynamics*. Springer, Switzerland

Klyashtorin LB (2001) Climate change and long-term fluctuations of commercial catches—the possibility of forecasting. Fisheries Technical Paper 410, FAO, Rome

Larkin PA (1977) An epitaph for the concept of maximum sustained yield. *Transactions of the American Fisheries Society* 106:1–11

Lett P, Mohn R, Gray D (1981) Density-dependent processes and management strategy for the Northwest Atlantic harp seal populations. In: CW F, TD S (eds) *Dynamics of Large Mammal Populations*, Wiley, New York, pp 135–158

Levi D, Giannetti G (1973) Fuel consumption as an index of fishing effort. In *General Fisheries Commission for the Mediterranean*. FAO, pp 1–17

Luce R, Raiffa H (1967) *Games and Decision*. Wiley, New York

Murphy G (1966) Population biology of the pacific sardine (*Sardinops caerulea*). Proceedings of the California Academy of Science 34:1–84

Pearse P (1967) The optimal forest rotation. *The Forestry Chronicle* 43:178–195

Reed W (1980) Optimum age-specific harvesting in a nonlinear population model. *Biometrics* 36:579–593

Schaefer MB (1967) Fishery dynamics and the present status of the yellowfin tuna population of the Eastern Pacific Ocean. *Bulletin of the Inter-American Tropical Tuna Commission* 2:247–285

Walters C (1975) Optimal harvest strategies far salmon in relation to environmental variability and uncertainty about production parameters. In *Workshop on Salmon Management*. Vancouver, B.C

Watt K (1968) *Ecology and Resource Management*. McGraw-Hill, New York

Chapter 9
Problems on the Management of Renewable Resource Harvesting

Problem M1

The common pheasant (*Phasianus colchicus*) was introduced into Protection Island in 1937. By following the demographic growth of this bird along time it has been possible to obtain the relationship linking the population rate of increase to the population size, as shown in the figure.

Assume that pheasant hunting is permitted and the regulation policy is based on granting licenses. You know that

- effort is measured as No. of operating hunters
- the catchability coefficient is 0.01 No. hunters^{-1} year^{-1}
- only half of licensed hunters is actually hunting in the average.

Calculate the number L of licenses to be granted that guarantees the maximum sustainable yield and the corresponding pheasant population size at equilibrium. Finally calculate the effort E_{ext} that would lead population to extinction.

Problem M2

The graph shown in the figure reports the stock-recruitment relationship for the sockeye salmon (*Oncorhynchus nerka*) of river Skeena (British Columbia, Canada). Approximately determine the maximum sustainable yield (as million fish).

Problem M3

The northern prawn fishery is one of the most important Australian fisheries. The tiger prawns *Penaeus esculentus* and *Penaeus semisulcatus* are two most prominent

M. Gatto and R. Casagrandi, *Ecosystem Conservation and Management*,
https://doi.org/10.1007/978-3-031-09480-4_9

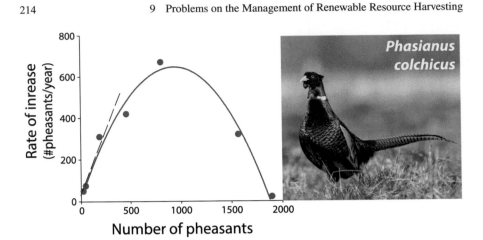

Problem M1 The rate of increase of the common pheasant (*Phasianus colchicus*) in Protection Island

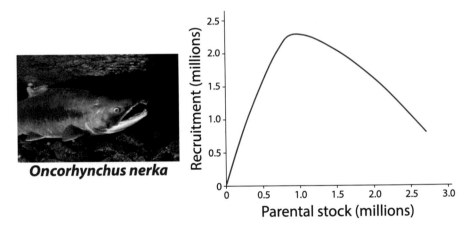

Problem M2 The stock-recruitment relationship for sockeye salmon (*Oncorhynchus nerka*) in river Skeena

species subject to harvesting. Wang and Die (1996) determined the stock-recruitment curves for the two species from available data.

If B_k is the total biomass (tonnes) in year k, we can approximate the curves by a Beverton-Holt model as follows:

$$B_{k+1} = \lambda B_k / (1 + \alpha B_k)$$

where

- for *Penaeus esculentus*: $\lambda = 1.6$ and $\alpha = 0.00019$;
- for *Penaeus semisulcatus*: $\lambda = 2.16$ and $\alpha = 0.0004$.

The effort is measured as boat-days (that is the number of operating boats times the days they are operating in a given year). From the estimates of the two researchers we can derive the catchability coefficients for the two prawns:

- *Penaeus esculentus*: $q = 3 \times 10^{-4}$ (boat-days)$^{-1}$;
- *Penaeus semisulcatus*: $q = 3.5 \times 10^{-4}$ (boat-days)$^{-1}$.

For each of the two prawns, determine:

(a) the Maximum Sustainable Yield (MSY);
(b) the effort that provides the MSY;
(c) the corresponding prawn biomass.

Problem M4

Zhang et al. (2012) describe the dynamics of the Korean stock of Japanese horse mackerel (*Trachurus japonicus*). From data, as shown in the Figure, they have obtained a stock-recruitment relationship linking the biomass (in metric tons) of one generation (the spawners B_t) to the biomass of the subsequent generation (the recruitment B_{t+1}).

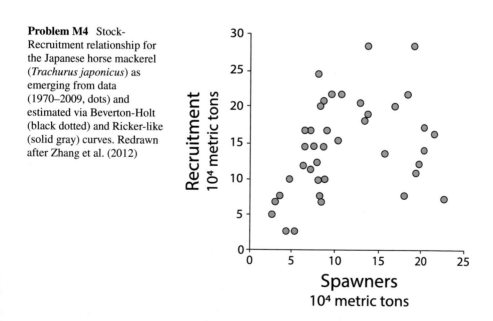

Problem M4 Stock-Recruitment relationship for the Japanese horse mackerel (*Trachurus japonicus*) as emerging from data (1970–2009, dots) and estimated via Beverton-Holt (black dotted) and Ricker-like (solid gray) curves. Redrawn after Zhang et al. (2012)

$$B_{t+1} = A(1 - \exp(-\lambda B_t))$$

with $A = 1.9 \times 10^5$ tonnes and $\lambda = 1.42 \times 10^{-5}$ tonnes^{-1}.

You are required to:

(a) Find the constant escapement policy that maximizes the sustainable yield of the mackerel stock;
(b) Calculate the corresponding MSY;
(c) Calculate the fraction u of the recruitment that is taken by means of the optimal policy at equilibrium.

Finally, assume that effort is measured in number of operating vessels, that the catchability coefficient is $q = 0.02$ (No. of vessels)$^{-1}$ year^{-1} and that there are 150 fishing vessels that operate during a fishing season of duration T (months); how long should T be for actually harvesting the fraction u?

Problem M5

Differently from Pacific salmon, the Atlantic salmon (*Salmo salar*) can reproduce several times. In other words the survival of reproductive adults is not zero.

For the small Norwegian Imsa River the relationship between the adults (measured as tonnes of biomass) returning for the first time to the river (A) in order to reproduce is approximately linked to the total biomass (tonnes) of adults (N) by the following relationship (Jonsson et al. 1998) between year k and year $k + 1$:

$$A_{k+1} = \frac{\lambda N_k}{(1 + \alpha N_k)}$$

Problem M5 The Atlantic salmon

Salmo salar

with $\lambda = 2500$ and $\alpha = 24$ tonne^{-1}. As some salmon survive after reproduction the total biomass of adults in year $k + 1$ is given by the sum of A_{k+1} plus the fraction s of adults N_k that survive to the next year. Assume $s = 0.25$.

Determine the constant escapement policy that provides the Maximum Sustainable Yield (measured as number of harvested salmon).

Problem M6

McCullough (1981) studied the dynamics of grizzly bears (*Ursus arctos*) in the Yellowstone Park. He reports the stock-recruitment curve shown in the Figure, which links the number P_k of adult bears in year k (namely, the individuals that are 4-years-old or older in year k) to the recruitment (R_k), namely the number of adult bears four years later.

Assume that the Park management allows hunting of adult grizzlies according to the following policy

$$H_k = \begin{cases} 0 & \text{if } R_k \leq 35 \\ 0.5(R_k - 35) & \text{if } R_k > 35 \end{cases}$$

so that a minimal recruitment of 35 adult bears is guaranteed. Evaluate the implications of implementing this policy on the population dynamics.

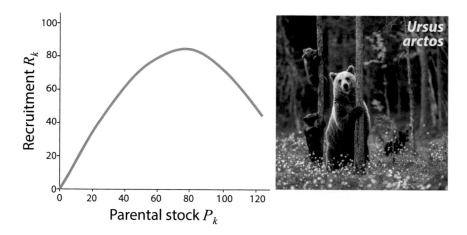

Problem M6 The stock-recruitment relationship for grizzly bears in the Yellowstone Park

Problem M7 The Atlantic
krill

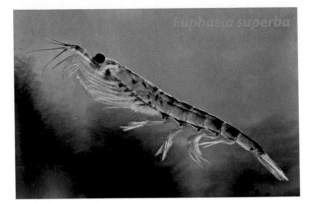

Problem M7

There has been an international debate in the past regarding the opportunity of catching Antarctic krill (*Euphausia superba*), a small crustacean that forms huge swarms in the waters surrounding Antarctica and is a fundamental food for many baleen whales. As the whale stocks have collapsed (and their recovery following the international moratorium on whale hunting will take decades), some researchers proposed that krill might be fished without impairing the functioning of the Antarctic ecosystem.

Discuss the problem by writing a simple prey-predator (krill biomass-whale biomass) of the Lotka-Volterra kind, in which the prey grows in a logistic way. Assume that krill is fished with a constant effort E, while hunting of whales does not take place. Determine the stable equilibria of the prey-predator system while the parameter E varies. Find out how the biomass of the sustainable krill catch varies for increasing E.

Problem M8

Assume you want to rationally regulate deer hunting in a grassland where the animals can extensively graze. To that end, describe the resource (grass)—consumer (deer) system dynamics by the equations

$$\frac{dG}{dt} = w - d_G G - pGD$$

$$\frac{dD}{dt} = -d_D D + epGD - uD$$

where G and D, respectively, indicate the biomass of grass and of deer (tonnes) and u is the mortality rate due to deer hunting. You know that

- w = net primary production = 100 tonnes year^{-1}
- d_G = grass death rate = 10 year^{-1}
- p = grazing rate coefficient = 0.2 tonnes^{-1} year^{-1}
- e = conversion coefficient = 0.1
- d_D = deer death rate = 0.1 year^{-1} .

Calculate the sustainable yield of deer biomass as a function of hunting mortality u. Then find the maximum sustainable yield.

Problem M9

Trawling is a major disturbance to fish habitat. In particular beds of *Posidonia oceanica*, an ecologically important seagrass, are badly impacted by this kind of fishing. The effect is noxious to the fishery itself because the fish carrying capacity (and thus the fish availability to the fishermen) is of course an increasing function of the remaining habitat. Thus if seagrass is destroyed the fishery yield decreases considerably.

To analyse the problem write a model that describes the dynamics of the fish stock biomass B (kg m^{-2}) and the *Posidonia* biomass P (kg m^{-2}). The dynamics of B is logistic, however the carrying capacity is not constant but increases with *Posidonia* biomass P. Fish biomass is harvested at a rate qEB where the effort E is measured as number of operating trawlers. The dynamics of *Posidonia* biomass P is described by a constant recruitment w of new seagrass biomass (kg m^{-2} year^{-1}) and a constant mortality rate m (year^{-1}). When the trawlers are operating, part of the *Posidonia* biomass is removed at a rate zEP. The parameter values are as follows:

- r = instantaneous rate of fish increase = 0.1 year^{-1};
- K = fish carrying capacity (kg m^{-2}) = kP where $k = 0.05$;
- $q = 0.01$ (No. of trawlers)$^{-1}$ year^{-1};

Problem M9 The habitat created by *Posidonia oceanica*

Posidonia oceanica

- $w = 0.9\,\mathrm{kg\ m^{-2}\ year^{-1}}$;
- $m = 1.8\,\mathrm{year^{-1}}$;
- $z = 0.05\,\mathrm{(No.\ of\ trawlers)^{-1}\ year^{-1}}$.

Find out the effort (number of trawlers) that maximizes the sustainable yield of fish biomass. Calculate the corresponding standing biomass of fish and *Posidonia* and the MSY.

Problem M10

The whales of the genus *Balaenoptera* have been severely depleted by the extensive hunting carried out during the twentieth century. A very rough way of measuring their total biomass is the Blue Whale Unit (BWU), adopted by the International Whaling Commission which equated two fin whales and six Sei whales to one blue whale. For the complex of these whales one can use a Schaefer model (logistic growth of whales and harvesting rate proportional to the product of effort and whale stock). The effort was measured as the total number of hunting days per year. Assume the following parameters

- K = carrying capacity = 400,000 BWU
- r = intrinsic instantaneous rate of increase = $0.05\,\mathrm{year^{-1}}$
- q = catchability coefficient = $1.3 \times 10^{-5}\,\mathrm{(hunting\ days)^{-1}}$

and calculate the bionomic equilibrium under the hypothesis that the opportunity cost of one hunting day is €5,000 and the selling price of one BWU is €75,000.

Problem M10 A 3D
rendering of a blue whale

Blue whale
(3D rendering)

Problem M11   The
European hake

Merluccius merluccius

Problem M11

The dynamics of the stock of the European hake (*Merluccius merluccius*) of the Adriatic sea can be described by the logistic model. Levi and Giannetti (1973) estimated the following demographic parameters:

- K = carrying capacity = 5,000 tonnes
- r = intrinsic instantaneous rate of increase = $1.7 \, \text{year}^{-1}$.

They also used the tonnes of consumed fuel per year as a measure of effort and estimated the catchability coefficient as $q = 2 \times 10^{-5} \, (\text{fuel tonne})^{-1}$.

Assume that the selling price of hake is €6,000 per hake tonne, while the opportunity cost is c = €140 per fuel tonne. Find the values of the hake stock and the effort at bionomic equilibrium. Then calculate the effort that would guarantee the maximum sustainable profit and the corresponding stock, catch and profit.

Problem M12

The Pacific halibut (*Hippoglossus stenolepis*) is a large bottom-feeding fish that is harvested in the North Pacific. The fishing vessels use gears called skates. A halibut skate is a long rope to which hooks are attached every ten yards or so. Effort can be thus measured as number of deployed skates.

Clark (1976) reports the following values for a logistic model describing the halibut population:

- intrinsic instantaneous rate of demographic increase $r = 0.71 \, \text{year}^{-1}$;
- carrying capacity $K = 80.5 \times 10^6$ kg.

Before 1930 the fishery was not regulated, so we may assume that the bionomic equilibrium had been reached, namely $N_B = 17.5 \times 10^6$ kg. It is also estimated that the catchability coefficient is $q = 10^{-3} \, (\text{No. of skates})^{-1} \, \text{year}^{-1}$.

Let c be the opportunity cost of operating one skate per year and p the selling price of one kg of halibut. From this information derive:

   (i) the ratio c/p,
  (ii) the optimal effort that maximizes the total profit of the fishery, and
 (iii) the corresponding annual yield in kg.

Problem M12   The Pacific
halibut

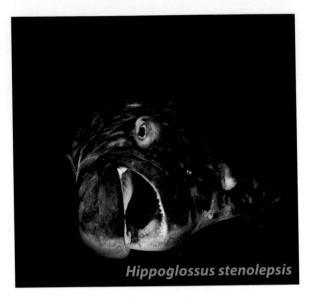

Hippoglossus stenolepsis

Problem M13

The dynamics of the fin whale (*Balaenoptera physalus*) is reasonably well described
by the following generalized logistic model (N is the whale numbers)

$$\frac{dN}{dt} = rN\left(1 - \frac{N}{K}\right)^a$$

Balaenoptera physalus

Problem M13   The fin whale

with $r = 0.06$ year^{-1}, $K = 400,000$ and $a = 0.143$. This stock is now protected, because it has been overexploited. Assume that, once the stock recovers (which will take several decades), hunting is permitted again, but is regulated. Implement a rational management scheme based on the information that the catchability coefficient is $q = 1.3 \times 10^{-5}$ (hunting days)$^{-1}$, the price of one whale is €40,000, the opportunity cost of one hunting day is €4,000.

Find the effort E_0 that maximizes the sustainable profit, and the corresponding profit. Suppose you want to utilize a tax τ (which is levied on each caught whale) as a regulation tool. Calibrate τ so that effort stabilizes to the same E_0 you have already calculated.

Problem M14

Consider an open-access renewable resource, which is simply regulated by imposing a tax on the net profit obtained by each economic operator exploiting the resource. Comment on the efficacy of such a regulation method by using H. S. Gordon's theory on open-access resources.

Problem M15

One of the main tenets of social welfare is progressive taxation. Assume that decision makers want to levy a progressive tax on the amount of biomass removed. The tax is structured as follows

$$\tau = \begin{cases} 0 & \text{if } Y \leq Y_0 \\ \tau_{\max} \dfrac{Y - Y_0}{Y} & \text{if } Y > Y_0 \end{cases}$$

where τ is the amount that is levied on each unit of biomass being harvested and Y_0 is the biomass yield below which the exploiters are exempted from paying taxes. Discuss the efficacy of this regulation method by using a Gordon-Schaefer model (logistic demography, harvest proportional to effort times biomass, and bionomic equilibrium).

Problem M16

In the forested areas of North America the beaver (*Castor canadensis*), which is no longer hunted for its fur, is becoming an important factor of nuisance to timber production. In fact, this herbivore can fell mature trees up to a diameter of 40 cm. A

Problem M16   The beaver

Castor canadensis

cost-benefit analysis estimated the annual damage of one beaver to be \$45. Assume that the local authority wants to implement a culling policy in an area where the beaver carrying capacity is 1 individual km^{-2} and the intrinsic instantaneous rate of demographic increase is 0.3 year^{-1}. Culling is expensive and the cost C can be evaluated as dollars per year per km^2. C can be assumed to be proportional to the inflicted mortality rate m [year^{-1}] and inversely proportional to the beaver density N [No. of beavers km^{-2}], because capturing beavers is harder when their density is lower. Specifically, assume $C = 2.5\ m/N$.

Find the best mortality rate, namely the one that minimizes the sum of the damages to timber plus the cost of culling. Estimate the beaver density at equilibrium corresponding to the optimal culling rate.

Problem M17

A tree species can be harvested for producing either pulpwood or sawlogs. The figure reports the net profit that can be obtained from clear-cutting a forest lot at different ages. Assume that the discount rate is zero and find out the optimal rotation periods for the two alternatives. Evaluate whether it is more convenient to produce pulpwood or sawlogs.

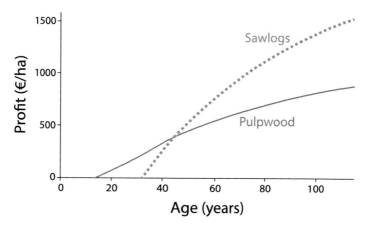

Problem M17   Net profit per hectare as a function of age for a tree species that can be harvested for producing pulpwood or sawlogs

Problem M18

The loblolly pine (*Pinus taeda*) is a conifer that is harvested in North America for its sawwood. The biomass of merchantable timber is a function of loblolly age and can be assumed to be given by

$$B(T) = \begin{cases} 0 & \text{if } T < 6 \\ 9.18T - 0.077T^2 - 55.09 & \text{if } T \geq 6 \end{cases}$$

where T is age in years and B is the biomass (tonnes/acre) that can be harvested and sold. The selling price of one merchantable tonne of pine is about \$30. The yearly maintenance cost of one acre of forest is \$9, while the cost of clearcutting and replanting one acre is \$125 independently of the age of the trees.

Problem M18   The loblolly pine

Assume that the discount rate is zero and find the optimal rotation period of the loblolly pine plantation.

Problem M19

You want to manage farmed salmon in Norway. In particular you would like to know the optimal period at which you should harvest all the salmon that you initially stocked as baby salmon in a farming cage. The data are as follows:

- In each cage you stock 100 baby salmon and the cost of purchasing and stocking each baby is €0.8;
- The mortality is practically zero, so you can neglect it;
- The salmon grow approximately in a quadratic way and the weight w of each salmon (grams) as a function of age T (days) is $w = 15T/(1 + 0.001T)$;
- The daily cost of feeding and managing the cage is 0.011 €day^{-1};
- The selling price of an adult salmon is 6 €kg^{-1}.

Find the optimal age at which you should harvest the salmon and stock another 100 babies, assuming that the rate of discount is zero. Calculate the profit from the cage.

Problem M20

Many seal populations are protected by law and international treaties. Fishermen, however, have often complained that the increase of seal numbers damages their catches, because these marine mammals subtract fish biomass that, without increased protection, would actually end up in their nets. Investigate the problem with reference to a constant-recruitment fish population in which the weight w of each individual fish varies with age τ [years] according to the following law

$$w(\tau) = w_{\max} (1 - \exp(-k\tau))$$

where

$$w_{\max} = 10 \text{ kg}$$
$$k = 0.1 \text{ year}^{-1}$$

while survival from age 0 (recruitment age) up to age τ follows the following law

$$p(x) = \exp(-\mu\tau)$$

with μ being a constant. Assume that cost of fishing effort is negligible and that the mortality rate without seals is $\mu_0 = 0.1\,year^{-1}$, while the mortality rate with seals is $\mu_S = 0.2\,year^{-1}$. Calculate the optimal age to be selected by the fishing gear in the two cases by assuming that the fishermen optimize the sustained biomass yield. Calculate the variation of yield between the two cases.

Problem M21

You have been requested to manage aquaculture in the lagoon of Stillwater. Every year, juveniles of the fantasy species *Argenteus bonissimus*, which is very much appreciated by gourmets, are recruited from the open sea to the lagoon. The weight w [kg] of the fish increases with age τ [years] according to

$$w(\tau) = w_{\max}\,(1 - \exp(-k\tau))$$

where $w_{\max} = 2\,kg$ and k is a growth coefficient which depends on the food provided to *A. bonissimus*. It equals $0.25\,year^{-1}$ if a highly caloric fish-meal C_1 is fed to the fish or $0.15\,year^{-1}$ if a poorer food C_2 is employed. The mortality rate of the fish is constant with age and equal to $0.1\,year^{-1}$. Also, there is no reproduction in the lagoon and the recruitment is constant and equal to 100,000 juveniles per year. The following economic data are available:

(i) the selling price of 1 kg of *A. bonissimus* is €10;
(ii) the harvesting cost is negligible whatever the harvested biomass and the mesh of the gear;
(iii) the annual cost of using C_1 is €400,000 while that of C_2 is €150,000.

   Determine the best harvesting policy (selected age) and the optimal fish-meal.

Problem M22

The spiny lobster (*Palinurus elephas*) is widespread in western Mediterranean and is an important resource for Sardinian fisheries. Bevacqua et al. (2010) and Tidu et al. (2004) have determined the following relationship linking the age x (years) to the fresh weight w (g) of lobsters in Sardinia:

$$w(x) = w_{\max}(1 - \exp(-kx))^3$$

with $w_{max} = 1300\,g$ and $k = 0.16\,year^{-1}$.
   Also, Bevacqua et al. (2010) estimated that the mortality rate μ of lobsters is constant and equal to $0.27\,year^{-1}$. You want to manage a stock of lobsters for which the recruitment to age 0 is constant and equal to 10,000 young lobsters. Suppose that

Problem M22 The spiny lobster

Palinurus elephas

the cost of fishing effort is negligible whatever the harvested biomass and the mesh of the fishing gear. The selling price of 1 kg of lobsters is €60. Determine

(a) the best age to be selected by the fishing year,
(b) the corresponding harvested biomass, and the corresponding economic return.

References

Bevacqua D, Melià P, Follesa MC, De Leo GA, Gatto M, Cau A (2010) Body growth and mortality of the spiny lobster *Palinurus elephas* within and outside a small marine protected area. *Fisheries Research* 106:543–549

Clark C (1976) *Mathematical Bioeconomics*. Wiley, New York

Jonsson N, Jonsson B, Hansen LP (1998) The relative role of density-dependent and density-independent survival in the life cycle of atlantic salmon *Salmo salar*. *Journal of Animal Ecology* 67:751–762

Levi D, Giannetti G (1973) Fuel consumption as an index of fishing effort. In *General Fisheries Commission for the Mediterranean*. FAO, pp 1–17

McCullough DR (1981) Population dynamics of the Yellowstone grizzly bear. In: Fowler CW, Smith TD (eds) *Dynamics of Large Mammal Populations*. Wiley, New York, pp 173–196

Tidu C, Sardà R, Pinna M, Cannas A, Meloni M, Lecca E, Savarino R, (2004) Morphometric relationships of the European spiny lobster *Palinurus elephas* from northwestern Sardinia. *Fisheries Research* 69:371–379

Wang YG, Die D (1996) Stock-recruitment relationships of the tiger prawns (*Penaeus esculentus* and *Penaeus semisulcatus*) in the Australian northern prawn fishery. *Marine and Freshwater Research* 47:87–95

Zhang CI, Lee JB, Lee DW, Choi I, (2012) Forecasting biomass and recruits by age-structured spawner-recruit model incorporating environmental variables. *Journal of the Korean Society of Fisheries and Ocean Technology* 48:445–451

Part IV
Parasite and Disease Ecology

Chapter 10
Ecology of Parasites and Infectious Diseases

10.1 Introduction

Only recently have ecologists begun to recognize the fundamental importance of disease and parasites in animal and plant population dynamics (Scott and Dobson 1989). The study of disease dynamics is very old (Bernoulli 1760), but it remained confined for a long time in the epidemiological literature, with special reference to the human species and domestic animals. The traditional view in ecology was that parasites, especially if they were adapted since long time to their environment, did not have a very important impact on the demography of their hosts. In contrast, during the past forty years, starting from the fundamental works by Anderson and May (1978, 1979) and May and Anderson (1979), scientists have understood that parasites are a crucial component of biodiversity (Hudson et al. 2006), whose impacts have direct and indirect effects on all species in all communities and ecosystems.

In some cases interactions are subtle and difficult to detect, but in other cases the effects of parasites are very clear: the introduction or loss of a single pathogen can completely alter the structure and dynamics of an ecological community. For example, the massive mortality originating from some epidemic cycles can control the demography of hosts much more effectively than intra or inter-specific competition (Dobson and MacCallum 1997). One of the best documented cases is that of red grouse, *Lagopus lagopus scoticus*, in northern England (see Fig. 10.1) and Scotland. Hudson et al. (1992) and Dobson and Hudson (1992) have clearly showed that the density of *Trichostrongylus tenuis* nematodes is a determinant factor that influences the size of the brood of grouse, the juvenile survival and the risk of predation by foxes.

In addition, endemic parasites can act as keystone species which can affect the functioning and the diversity of ecosystems. For instance, when the rinderpest virus spread in East Africa at the end of 1800 (Dobson 1988), there has been a collapse in the abundance of wild herbivores (wildebeests, zebras, buffalo, etc.), which are characterized by an amazing number of species, varieties of ecological roles and ecological niches. The collapse had important consequences on inter-specific inter-

© The Author(s), under exclusive license to Springer Nature Switzerland AG 2022
M. Gatto and R. Casagrandi, *Ecosystem Conservation and Management*,
https://doi.org/10.1007/978-3-031-09480-4_10

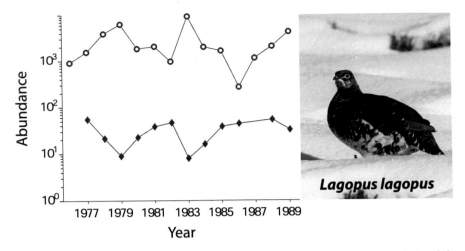

Fig. 10.1 Trend in the number of red grouse (*Lagopus lagopus scoticus*, white circles) and the average load of the parasite *Trichostrongylus tenuis* (No. of nematodes per grouse, dark diamonds) in the site of Gunnerside (Northern England), redrawn after Dobson and Hudson (1992). Note that data are plotted on a logarithmic scale

actions, on the vegetation structure and on fire regimes, as documented by studies on the famous Serengeti National Park (Tanzania). The subsequent recovery of herbivores after cattle vaccination practised around 1950 (Scott 1964; Plowright 1985) demonstrates that the absence or presence of this disease does regulate the ecosystem functioning. In contrast to what was previously thought, thus, the dynamics of African herbivores is not regulated by large predators only. On the contrary, the opposite may be true: the predators are tracking the dynamics of their prey, which is controlled by pathogens like the rinderpest virus.

A separate discussion is deserved by some special organisms that parasitize animals, whose effects, however, are more similar to those of predators: *parasitoids*. They are usually insects (of the order *Hymenoptera*) living in the wild at the adult stage, which lay their eggs on or inside other insects; the larvae of parasitoids develop by consuming, and then killing, their host. In contrast to proper parasites, whose body sizes are much smaller than those of the host organism, parasitoids have a relatively large body, reaching sometimes almost the same size of the host. An example of host-parasitoid dynamics is reported in Fig. 10.2 which shows the coupled oscillations of a population of the wasp *Heterospilus prosopidis* and of its host (*Callosobruchus chinensis*). The parasitoids play an important role from the viewpoint of applications, because they are often used for *biological control* as an effective substitute for pesticides. Many insects, or more in general harmful organisms, in fact turn out to be potential hosts of parasitoids, which are then released into the environment to clear crops from pest.

A good part of modern experimental and field studies on the ecology of parasites is based on the theoretical work of ecologists who have highlighted the fundamental

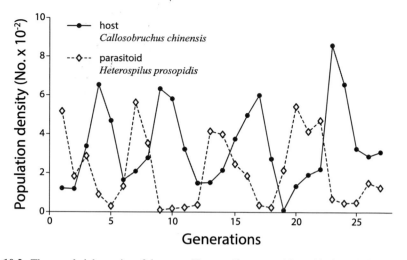

Fig. 10.2 The coupled dynamics of the wasp *Heterospilus prosopidis* and its host *Callosobruchus chinensis*, the azuki bean weevil, a dangerous insect pest. Data are from population "E" of the famous experiments by Utida (1957)

role of parasitism in controlling animal and plant populations. Before the 1970s, most theoretical studies in epidemiology were aimed at analysing the disease dynamics in host populations that are basically constant over time. This approximation is valid for many diseases affecting human populations (see the book by Anderson and May 1991), but is unreasonable for the dynamics of many natural populations of animals and plants. Anderson and May (1979) and May and Anderson (1979) proposed a unified paradigm for the study of the influence of diseases on the dynamics of their hosts. They divided the parasites in two large groups: the *microparasites* (mainly bacteria and viruses, see Fig. 10.3a) and *macroparasites* (essentially helminth worms and arthropods, see Fig. 10.3b). The first reproduce within the host, often within individual cells, and have a life cycle shorter than the average life-time of the host. The latter grow within the host but reproduce via infective stages that are released outside of it. Also, they have a relatively long average life-time, which is comparable to that of the host. These different characteristics of parasites have an influence on the type of models that are used in the two cases. The dynamics of microparasites abundance can be neglected and thus the disease dynamics can be described by taking into account only the changes occurring in the host population, suitably classified according to their state of infection (typical classes of hosts are the susceptible to the disease, the infected and the recovered). For macroparasites it is instead necessary to take into account the dynamics of the numbers of parasites themselves and the parasite load in their hosts. In fact, their impact in terms of mortality, morbidity or reduced fertility depends on the number of parasites infecting the host.

The modes of infection between hosts can be very different depending on the parasite considered. Table 10.1 provides a rough classification concerning micropar- asite diseases. Rarely, transmission is *vertical*, i.e. it occurs from the mother to the

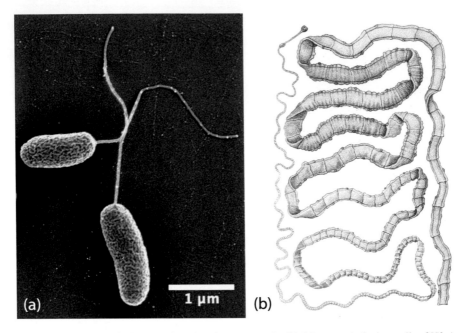

Fig. 10.3 Examples of microparasite (**a**) and macroparasite (**b**). More precisely, in **a** cells of *Vibrio cholerae*, the agent of cholera (Waldor and Ray Chaudhuri 2000), are imaged by scanning electron microscopy within the H2020 project CMDNAUP; in **b** the pig tapeworm *Taenia solium* is sketched in an ancient drawing of the XIX century. Note that the lenght of an adult *Taenia solium* is on the order of a few meters

embryo or to the newborn, more often being *horizontal*. This can take place through *direct contact*, through a biotic *vector* (e.g. the *Anopheles* mosquito for malaria), via *water*, or by means of pathogens that persist in the environment (e.g. in the form of *spores*).

Macroparasites are often large in size at their adult stage, in particular helminth worms such as nematodes (e.g. roundworms and protostrongylids) and cestodes (e.g. tapeworms). They may have a simple life-cycle (see the cycle of *Ascaris suum* in Fig. 10.4a): adult parasites reproduce within a given host that releases outside the propagules, typically eggs, which are then either directly re-ingested by another host of the same species or develop into successive stages (e.g. larvae) before being ingested. In other cases, macroparasites do need not only a definitive host for developing into an adult stage but also an intermediate host inside which they develop particular juvenile stages that are then ingested by the final host (see, for example, the cycle of *Fasciolopsis buski* in Fig. 10.4b). Even more complex life cycles are possible, yet they are not described herein.

Despite the incredible advances in medicine, infectious diseases are still a leading cause of death for humans, representing more than 25% of annual deaths (Morens et al. 2004). Some of these diseases were believed to be definitively defeated (such as tuberculosis or cholera), but instead re-emerged in often drug-resistant variants.

Table 10.1 Different modes of disease transmission in microparasites

Transmission mode	Description	Examples
Direct	The disease propagules are transmitted directly from one host to another through air or by physical contact (including sexual intercourse)	Common cold, measles, syphilis, HIV/AIDS, rabies, COVID-19
Vector borne	Propagules are transported from one host to another by means of a secondary host species, the vector (e.g. a mosquito)	Malaria (*Anopheles* mosquito), dengue (tiger mosquito)
Water borne	Propagules are transmitted through contaminated water	Cholera, rotavirus
Environmental	Propagules are released into the environment where they remain until they are acquired by another host. They differ from directly transmitted pathogens, because of their greater capacity to persist in the external environment	Smallpox, tetanus, anthrax
Vertical	Propagules are passed from mother to offspring through body fluids or milk	HIV/AIDS, hepatitis B and C

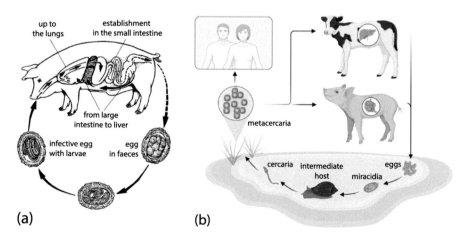

(a)   (b)

Fig. 10.4 Examples of macroparasite life-cycles. **a** Simple cycle of a nematode (*Ascaris suum*) (modified after Roepstorff and Nansen 1998); **b** Cycle with an intermediate host of *Fasciolopsis buski* (modified after Siles-Lucas et al. 2021)

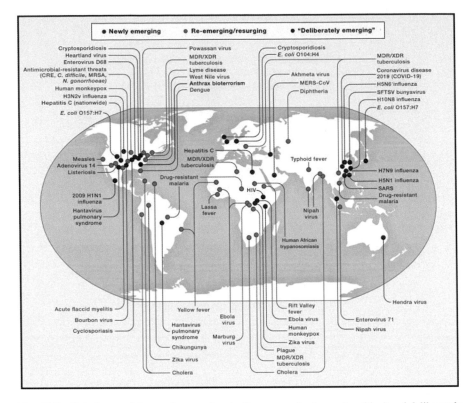

Fig. 10.5 Global map of the newly emerging (red), re-emerging/resurging (blue) and deliberately emerging (black) diseases. Taken from Morens and Fauci (2020)

Others are new diseases emerged in the past fifty years, such as AIDS/HIV, Ebola haemorrhagic fever, acute respiratory syndrome SARS, or COVID-19 (Morens and Fauci 2020). Figure 10.5 shows the global map of emerging or re-emerging diseases. Jones et al. (2008) have compiled a database of infectious diseases that emerged between 1940 and 2004, recording 335 events. The number of deaths linked to them has been increasing since 1940, reaching a peak between 1980 and 1990 (see Fig. 10.6) in conjunction with the AIDS pandemic. More recently, this peak has been surpassed by the casualties caused by COVID-19. As noted by Jones et al. (2008) and Lipkin (2013), the majority (more than 60%) of infectious events are *zoonoses* (i.e. diseases related to pathogens of non-human animal origin, see Fig. 10.7) and most of these zoonoses (more than 70%) are due to pathogens whose reservoir is wildlife. This shows, as a matter of fact, that there is a *continuum* that links wildlife, domestic animals and humans (Daszak et al. 2000), with an exchange of pathogens between compartments. Therefore, it is now clearer and clearer that parasite ecology is not just a chapter of ecology that is basic to understanding ecosystems functioning, but it is also a fundamental pillar for designing policies of public health protection. The traditional approach to the treatment of infectious diseases based on pharmacolog-

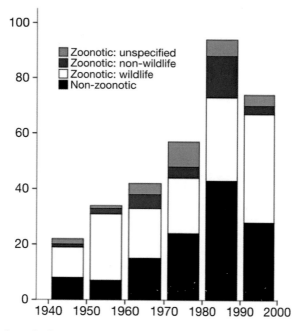

Fig. 10.6 Global trend of new emerging infectious diseases subdivided according to pathogen reservoir. Taken from Jones et al. (2008)

ical treatment must necessarily be complemented by both an understanding of the natural environment of which man is part, and rational protection and management of ecosystems including the compartments of parasites and pathogens.

10.2 Dynamics of Diseases Caused by Microparasites

The essential feature of microparasites is that they reproduce within the host and have a life cycle shorter than the host average lifetime. Therefore, we can capture the dynamics of many diseases through a compartmental model in which hosts are divided in several categories with respect to their ability to be infected and to infect. Depending on the transmission mode, the compartments can be of various kinds. For directly transmitted diseases (human-to-human, animal-to-animal) it is sufficient to account for a single host species divided into several compartments. In water-borne diseases the microparasite can be hosted in both humans (or animals) and water, which is thus contaminated and becomes a source of infection for the hosts. In vector-borne diseases the pathogen needs two species for completing its life cycle: some stages are hosted by a vector species and the infection of the focal host is never direct, but is due to the encounter with a vector; non-infected vectors become carriers of the pathogen via the encounter with an infected host.

We will separately deal with the three categories of microparasitic diseases.

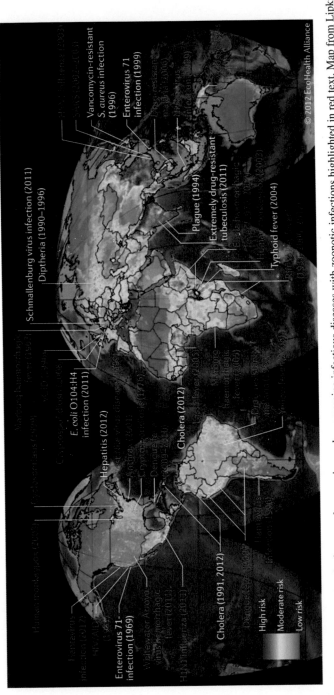

Fig. 10.7 Hotspots of outbreaks for recently emerging and re-emerging infectious diseases with zoonotic infections highlighted in red text. Map from Lipkin (2013)

10.2.1 Directly Transmitted Diseases

This kind of diseases caused by microparasites can be divided into two major classes: there are diseases that, once contracted, leave recovered individuals in a semi-permanent or permanent immunization state. There are instead diseases that can be contracted several times by the same individual. Examples of the first class are measles and chickenpox, examples of the second are influenza, common cold and pertussis. Actually, this subdivision is not so sharp in reality: for the same disease, the degree of immunization can vary from individual to individual and from species to species. In diseases without any immunization, individuals are usually divided into *susceptible* (indicated with S) and *infected* (and infectious, indicated with I). The recovered from the disease become immediately susceptible again. Models describing diseases of this kind are therefore called SI. In diseases with partial or complete immunization, a third category must be introduced, that of *recovered* and immune (indicated by the letter R). After a certain time, the recovered can lose their immunization and become susceptible again. In this case, the models are called SIR. In the following, we mainly introduce SI models, which are simpler, though less realistic, and then illustrate some results concerning SIR models.

In addition, there are diseases with a long incubation period during which organisms are infected, but are still unable to transmit the disease. In these cases, it is necessary to introduce a further infection class: that of the *exposed* (i.e. infected but not yet infectious, indicated with E). It is then also possible to have models of SEI or SEIR type. For simplicity, here we will not discuss these two types of models. For those who want to better investigate the matter, Heesterbeek and Roberts (1995) and Iannelli and Pugliese (2014) provide a broad overview of the different microparasite models.

10.2.1.1 SI Models

The dynamics of infection is fully characterized by the temporal changes of the variables S and I, i.e. the density (measured for instance as number of organisms per km^2) of the susceptibles and the infecteds. A general formulation of a model with continuous-time reproduction and no immunization is the following:

$$\frac{dS}{dt} = v_S S + v_I I - \mu S - iS + \gamma I$$

$$\frac{dI}{dt} = iS - (\mu + \alpha + \gamma) I \tag{10.1}$$

where t represents time (measured in appropriate units, e.g. years), ν_S and ν_I are the birth rates of susceptibles and infecteds respectively, μ is the mortality rate in the absence of disease, i is the rate at which the susceptible are infected, α is the mortality rate caused by the disease (sometimes used as an indicator of disease *virulence*) and γ is the recovery rate. The birth rate of the infected ν_I is obviously lower than that of the susceptible (ν_S). A hypothesis that is often reasonable and that we will we use here is that this rate is negligible ($\nu_I = 0$), namely that organisms do not reproduce during the course of the disease.

It is important to introduce some terms used in epidemiology. The flow of new infected iS is called the *incidence* rate of the disease (measured in the number of new infected per unit time). The ratio I/N, with $N = S + I = $ total density of organisms, is called *prevalence* of the disease and is nothing but the fraction of the population that is sick, and therefore infected and infectious.

Model (10.1) takes different forms and has different dynamic behaviours depending on how one specifies the various demographic and epidemiological parameters that characterize it. The first important specification concerns the population demography in absence of infection. The two most common assumptions are that hosts follow either a Malthusian dynamics or a logistic one. In the first case, the rates of birth (ν_S) and death (μ) are constant; in the second case, the per capita growth rate of the disease-free population $\nu_S - \mu$ is a linear and decreasing function of density N. The second specification concerns the incidence rate of the disease. The parameter i represents the probability per unit time that a susceptible becomes infected; it is given by the product of three factors:

(A) the probability $c(N)$ that a susceptible contacts another organism in a time unit,
(B) the probability I/N (prevalence) that this other organism be infected,
(C) the probability of actually becoming infected and infectious, which from now on we suppose to be constant.

Depending on the assumptions on the contact rate $c(N)$, several models of disease transmission can be obtained. In particular, we can make two extreme hypotheses:

(1) the number of contacts per unit time is proportional to the density N of organisms, for example because the disease is airborne, such as influenza;
(2) the number of contacts per unit time is constant, for example because the disease is sexually transmitted and the number of sexual contacts depends on the behaviour of each organism, not on population density.

In case (1) the rate of infection i is proportional to $N \times I/N$, thus ultimately to the density of infected I (law of mass action or *density-dependent transmission*). In case (2), the infection rate is proportional to I/N or the frequency of infected (*frequency-dependent transmission*). Actually, both assumptions are unrealistic: density-dependent transmission is unrealistic for very high values of density N, because it is unthinkable that the number of contacts that each organism has per unit time does not anyway saturate to a maximum value, while the frequency-dependent transmission is unrealistic for very low values of N, because it is unthinkable that in such a case the number of contacts per unit time remains constant (it must necessarily tend to zero for N tending to zero). A possible functional form of $c(N)$ reconciling assumptions (1) and (2) is the one we already mentioned in the case of predator's functional response (Sect. 2.4), namely $c(N)$ increases with N, but saturates to a maximum value. Let us have a look now at the consequences of the different assumptions.

10.2.1.2 Malthusian Demography and Density-Dependent Transmission

Since the infection rate i is proportional to the density of the infected, the SI model assumes in this case the following form

$$\dot{S} = rS - \beta IS + \gamma I$$
$$\dot{I} = \beta IS - (\mu + \alpha + \gamma) I \tag{10.2}$$

where $r = \nu_S - \mu$ is the instantaneous rate of Malthusian population growth and β is the coefficient of disease transmission from infected to susceptible (measured in time^{-1} number of infected^{-1}). Suppose now that the population can grow exponentially in the absence of the disease ($r > 0$). We show that the microparasite is able to regulate the population, that is to prevent the host population from exhibiting Malthusian growth. To this end, we note that model (10.2) has two equilibria states, obtained by setting to zero both time derivatives:

(i) $\bar{S}_0 = 0$, $\bar{I}_0 = 0$;

(ii) $\bar{S} = \dfrac{\mu + \alpha + \gamma}{\beta}$, $\bar{I} = \dfrac{r(\mu + \alpha + \gamma)}{\beta(\mu + \alpha)}$.

Like in the Malthusian model without disease, the first equilibrium is unstable, but the real novelty is represented by the presence of a second equilibrium in which the total number of organisms is finite and given by

$$\bar{N} = \bar{S} + \bar{I} = \frac{(\mu + \alpha + \gamma)(\nu_S + \alpha)}{\beta(\mu + \alpha)}.$$

It can be proved (for example by using the method of linearization) that this equilibrium is stable. Figure 10.8 shows the isoclines for model (10.2) and the evolution of some trajectories. The only trajectory that diverges exponentially is the one with initial conditions $I = 0$; all the others converge toward the stable equilibrium, thus demonstrating the effectiveness of the microparasite disease as demographic regulator. Note that the regulation effect is much more effective (lower $\bar{N}$) if the transmission coefficient β is greater, and it is much less effective (higher $\bar{N}$) if the recovery rate γ is larger, because this allows the replenishment of the susceptible class. However, high recovery rates guarantee the absence of oscillations in the dynamics toward equilibrium. Particularly interesting is the dependence of $\bar{N}$ from the disease virulence (mortality rate α). Indeed, the population size at the equilibrium appears to be either growing for all α's or decreasing for small α's and increasing for large α's, with a minimum value for intermediate virulence. Therefore, very virulent diseases do not regulate at all the population growth, as one might think. This becomes clearer by calculating the prevalence at equilibrium. It is given by

$$\frac{\bar{I}}{\bar{N}} = \frac{r}{v_S + \alpha}.$$

So, the larger is the disease virulence, the smaller is the fraction of infecteds in the total population. Basically, for very virulent diseases the reservoir of the infected is very small because infected organisms die very quickly. Fortunately, terrible diseases such as the haemorrhagic fever caused by the Ebola virus, cannot spread very effectively.

10.2.1.3 Logistic Demography and Density-Dependent Transmission

For the sake of simplicity, let us assume that the mortality rate μ is constant and that logistic demography is exclusively determined by the birth rate being dependent on density. The model SI then becomes

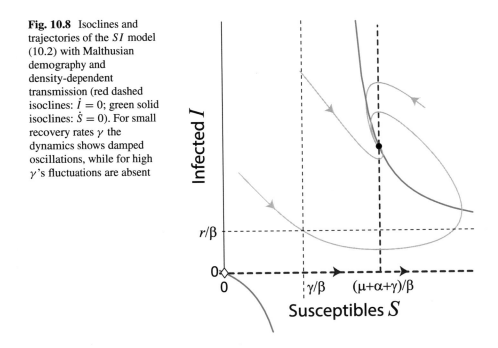

Fig. 10.8 Isoclines and trajectories of the SI model (10.2) with Malthusian demography and density-dependent transmission (red dashed isoclines: $\dot{I} = 0$; green solid isoclines: $\dot{S} = 0$). For small recovery rates γ the dynamics shows damped oscillations, while for high γ's fluctuations are absent

$$\dot{S} = rS\left(1 - \tfrac{S+I}{K}\right) - \beta IS + \gamma I$$
$$\dot{I} = \beta IS - (\mu + \alpha + \gamma)I \qquad (10.3)$$

where K is the population's carrying capacity in absence of infection. It is easy to find out that, under certain conditions (specified below), model (10.3) has three equilibria: the first is the trivial equilibrium $\bar{X}_0 = \left[\bar{S}_0 = 0, \bar{I}_0 = 0\right]^T$ (where the superscript T indicates matrix transposition); the second is the equilibrium corresponding to the healthy population at its carrying capacity $\left(\bar{X}_K = \left[\bar{S}_K = K, \bar{I}_K = 0\right]^T\right)$; the third equilibrium $\bar{X}_+$ corresponds to a partially infected population. In fact, it can be easily found that both time derivatives of model (10.3) vanish if

$$\bar{S}_+ = \frac{\mu + \alpha + \gamma}{\beta},$$
$$\bar{I}_+ = \frac{rK(\mu + \alpha + \gamma)}{\beta K(\mu + \alpha) + r(\mu + \alpha + \gamma)}\left(1 - \frac{\mu + \alpha + \gamma}{\beta K}\right). \qquad (10.4)$$

For this equilibrium to make sense in biological terms, it is required that $\bar{I}_+$ be not negative, i.e. that the following inequality holds

$$R_0 = \frac{\beta K}{\mu + \alpha + \gamma} \geq 1. \qquad (10.5)$$

The quantity R_0 is termed the *basic reproduction number* of the disease and is one of the key parameters of epidemiology. Its meaning is very similar to the net reproductive rate used in demography where R_0 is the average number of daughters produced by one female in her lifetime. In fact, since $1/(\mu + \alpha + \gamma)$ is the mean residence time in the infected class, R_0 can be interpreted as the average number of secondary infections (daughters) produced by one infected individual (primary infection, mother) introduced in a healthy population consisting of K individuals.

If R_0 is less than 1, the equilibrium with a partially infected population cannot exist. It can be shown that for $R_0 \leq 1$ the equilibrium $\bar{X}_K$ is stable whatever the initial conditions and so the disease spontaneously fades out. On the contrary, if R_0 is greater than unity, the equilibrium $\bar{X}_+$ (see Eq. 10.4) is the only stable equilibrium and therefore the microparasitic disease is endemic.

Figure 10.9 shows the isoclines of model (10.3) and the evolution of the trajectories in the latter case. It is interesting to notice that the larger is the carrying capacity of the host population or the larger is the transmission rate β, the larger is R_0. On the contrary, the larger is the recovery rate γ or the greater is the virulence α, the smaller is R_0. Once again, we remark that very lethal diseases are not as dangerous as one may think because they cannot spread very easily or they cannot even spread ($R_0 < 1$).

It is also interesting to calculate the prevalence at the equilibrium. It is given by

Fig. 10.9 Isoclines and trajectories of the SI model (10.3), with a logistic growth and density-dependent transmission (red dashed isocline: $\dot{I} = 0$; green solid isocline: $\dot{S} = 0$)

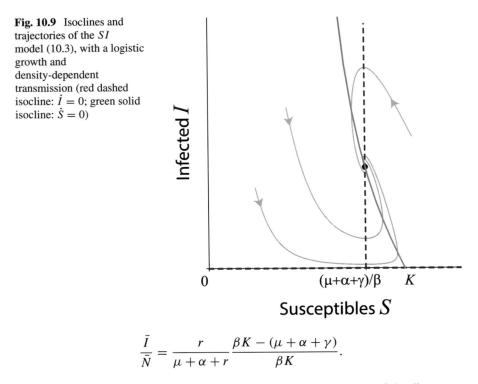

$$\frac{\bar{I}}{\bar{N}} = \frac{r}{\mu + \alpha + r} \frac{\beta K - (\mu + \alpha + \gamma)}{\beta K}.$$

The percentage of infected is then decreasing with the virulence α of the disease.

10.2.1.4   Logistic Demography and Saturating Transmission

If the contact rate is constant, then transmission is termed frequency-dependent. As mentioned above, the hypothesis that the number of contacts remains constant even for N tending to 0 is unrealistic. A hypothesis that couples density-dependent transmission for small N's and frequency-dependent transmission for intermediate or large N is one in which we assume that the contact rate $c(N)$ is increasing yet saturating with N. We can use an expression similar to that of the predator's functional response of the second type, namely $c(N) \propto N/(\delta + N)$, where δ is the half-saturation density. It follows that the infection rate i is given by

$$i = \beta \frac{N}{\delta + N} \frac{I}{N} = \beta \frac{I}{\delta + N}$$

where β is an appropriate coefficient of proportionality (measured as time^{-1}) which depends also on the probability of actually contracting the disease after having contacted an infected individual. The resulting SI model with logistic demography is

Fig. 10.10 Isoclines and trajectories of the SI model (10.7) with logistic demographics and saturating transmission. Unlike Figs. 10.8 and 10.9, the system dynamics is not represented in the SI state space (density of infected vs. density of susceptible) but in the state space $N - x$ (disease prevalence vs. total population density). Red dashed isocline: $\dot{x} = 0$; green solid isocline: $\dot{N} = 0$

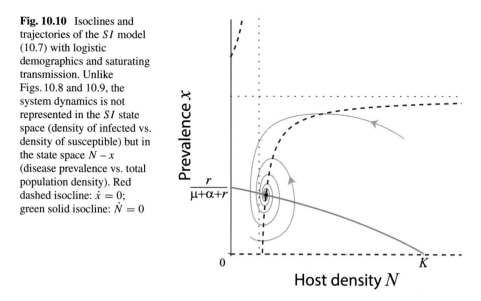

The equations (10.6):

$$\dot{S} = rS\left(1 - \frac{S+I}{K}\right) - \beta\frac{IS}{\delta+S+I} + \gamma I$$
$$\dot{I} = \beta\frac{IS}{\delta+S+I} - (\mu + \alpha + \gamma)I \tag{10.6}$$

The analysis of the model is easier after a change of variables. Instead of using S and I, it is convenient to introduce the total density of hosts $N = S + I$ and the prevalence $x = I/(S + I)$. Exploiting the relationships

$$\dot{N} = \dot{S} + \dot{I}$$
$$\dot{x} = \frac{1}{N}\dot{I} - \frac{I}{N^2}\dot{N}$$

one gets

$$\dot{N} = \left[r\left(1 - x\right)\left(1 - \frac{N}{K}\right) - (\mu + \alpha)x\right]N$$
$$\dot{x} = \left[\left(\beta\frac{N}{\delta+N} - r\left(1 - \frac{N}{K}\right) - (\mu + \alpha)\right)(1 - x) - \gamma\right]x \tag{10.7}$$

Like in the previous density-dependent case, this model has a trivial equilibrium $\bar{X}_0 = \left[\bar{N}_0 = 0, \bar{x}_0 = 0\right]^T$ and an equilibrium corresponding to the population being disease-free at its carrying capacity $\bar{X} = \left[\bar{N} = K, \bar{x} = 0\right]^T$. The non-trivial isoclines (see Fig. 10.10) have the following expressions

$$\dot{N} = 0 \quad \Rightarrow \quad x = \frac{r\left(1 - \frac{N}{K}\right)}{\mu + \alpha + r\left(1 - \frac{N}{K}\right)}$$

$$\dot{x} = 0 \quad \Rightarrow \quad x = 1 - \frac{\gamma}{\beta\frac{N}{\delta+N} - r\left(1 - \frac{N}{K}\right) - (\mu + \alpha)}$$

They are displayed in Fig. 10.10. It follows that there is a third equilibrium at which the population is partially infected only if the non-trivial isoclines intersect at values of N in the range $(0, K)$ and at values of prevalence x comprised between 0 and 1. It is easy to verify that the condition required for this to occur is

$$\beta \frac{K}{\delta + K} - (\mu + \alpha) > \gamma.$$

In this case, the third equilibrium is stable (see the trajectories shown in Fig. 10.10) and the trivial equilibria are unstable. If the above condition is not verified, the third equilibrium does not exist and the stable equilibrium is of course $\bar{X}_K$.

Even in this case we can define the basic reproduction number of the disease R_0, i.e. the average number of secondary infections produced by one infected individual introduced into a healthy population (which therefore consists of K individuals). As the number of new infections produced in a unit time is $iK = \beta \frac{1}{\delta+K} K$ and the mean residence time of the infected individual in the infected class is $1/(\mu + \alpha + \gamma)$, it follows that

$$R_0 = \frac{\beta K}{(\delta + K)(\mu + \alpha + \gamma)}.$$

It is then easy to verify that the condition for the existence of the third equilibrium is still nothing but the condition $R_0 > 1$. If the reproduction number of the disease is smaller than one, then, the disease cannot spread in the population. Note that for very small values of the half-saturation density δ, that is when the disease is practically frequency-dependent, one gets

$$R_0 \cong \frac{\beta}{\mu + \alpha + \gamma}.$$

Therefore, we can conclude that the spread of a frequency-dependent disease (e.g. AIDS) within a population is not facilitated by the size K of the population, in contrast to what has been found above for density-dependent transmission, where R_0 is directly proportional to the demographic carrying capacity of the population (see Eq. 10.5).

10.2.1.5   SIR Models

The dynamics of infection is fully characterized by the behaviour over time of the variables S, I and R, i.e. the density of susceptible, infected and recovered (and permanently or temporarily immune). A general formulation of a SIR model is as follows:

$$\begin{aligned}
\dot{S} &= \nu_S(S + R) + \nu_I I - \mu S - iS + \gamma R \\
\dot{I} &= iS - (\mu + \alpha + \rho)I \\
\dot{R} &= \rho I - (\mu + \gamma)R
\end{aligned} \tag{10.8}$$

where ν_S and ν_I are the birth rate of susceptibles and infecteds respectively, μ is the mortality rate in the absence of the disease, i is the rate at which susceptibles become infected, α is the mortality rate induced by the disease (virulence), ρ is the recovery rate resulting in immunization and γ is the rate of immunity loss. It is assumed that the recovered have the same birth and death rates of the susceptible and that their progeny does not inherit immunity and is therefore susceptible. Even in the case of SIR models, we make the assumption that the birth rate of the infected ν_I is negligible, i.e. that individuals do not reproduce during the disease course. The only case analysed here is the one with Malthusian dynamics and density-dependent transmission. Our purpose is in fact to investigate if a disease that provides complete or partial immunity is able to control an otherwise exponentially growing population.

The SIR model is the following

$$\begin{aligned}
\dot{S} &= rS + \nu_S R - \beta SI + \gamma R \\
\dot{I} &= \beta SI - (\mu + \alpha + \rho)I \\
\dot{R} &= \rho I - (\mu + \gamma)R
\end{aligned} \tag{10.9}$$

where $r = \nu_S - \mu > 0$ is the instantaneous Malthusian rate of population growth, β is the coefficient of disease transmission from infected to susceptible. Like in the corresponding SI model, there are two possible equilibria: the trivial equilibrium $S = 0$, $I = 0$, $R = 0$, which is always unstable because the population dynamics of the disease-free population is Malthusian, and a second equilibrium given by

$$\bar{X}_+ = \begin{bmatrix} \bar{S} = \dfrac{\mu + \alpha + \rho}{\beta} \\[2mm] \bar{I} = \dfrac{r(\mu + \alpha + \rho)(\mu + \gamma)}{\beta\left[(\mu + \alpha + \rho)(\mu + \gamma) - (\nu_S + \gamma)\rho\right]} \\[2mm] \bar{R} = \dfrac{r\rho(\mu + \alpha + \rho)}{\beta\left[(\mu + \alpha + \rho)(\mu + \gamma) - (\nu_S + \gamma)\rho\right]} \end{bmatrix}.$$

The constraint that $\bar{I}$ and $\bar{R}$ be positive, however, requires that the denominator of their expression be positive, namely that the recovery rate be such as to satisfy the inequality

$$\rho < \frac{(\mu + \alpha)(\mu + \gamma)}{r}. \tag{10.10}$$

It can be shown that, if the recovery rate is below the threshold given by (10.10), the population dynamics converges towards the equilibrium $\bar{X}_+$, otherwise the only equilibrium is the trivial one, which is unstable. We can therefore conclude that, in contrast to diseases that do not provide immunity, those that confer partial or complete immunization are able to control the population growth only if the recovery rate is not too high.

10.3 Water-Borne Diseases

Water-borne (WB) and water-related (WR) diseases are infections in which the causative agent (or its vector or host) spends at least part of its life cycle in water. A wide range of micro- (viruses, bacteria, protozoa) and macro-parasites (mostly flatworms and roundworms) is responsible for WB and WR infections, which are generally caused by exposure to (or ingestion of) water contaminated by pathogenic organisms. WB and WR diseases still represent a major threat to human health, especially in the developing world. As an example, diarrhoea, commonly associated with WB pathogens, is responsible for the deaths of about 1.5 million people every year, thus representing one of the leading causes of death among infants and children in low-income countries (World Health Organization 2014b). Most of that burden is attributable to unsafe water supply, lack of sanitation and poor hygienic conditions, which either directly or indirectly affect exposure and transmission rates (World Health Organization 2014a).

Here we focus on water-borne diseases that are related to micro-parasites. Perhaps, the best known of these diseases is cholera. It is caused by the bacterium *Vibrio cholerae* (Fig. 10.3a), which colonizes the human intestine and was discovered in 1854 by Pacini (1854) during the Florence epidemic. The transmission of the disease is mediated by water. In fact, *V. cholerae* is a natural member of the coastal aquatic microbial community and can survive outside the human host in the aquatic environment. Therefore, the disease can spread from the coastal region, where it is autochthonous, towards inland through waterways and river networks. The infection is always caused by ingestion of water either contaminated by *V. cholerae* present in a natural reservoir (primary route) or contaminated by humans (secondary infection), and thus the role of the aquatic environment is crucial for the transmission as well as for the spreading of the disease. Because of poor water sanitation, cholera still represents a global threat to public health, especially in developing countries (World Health Organization 2014a, b), as unfortunately shown by the catastrophic epidemic outbreak that has recently struck Haiti (e.g. Rinaldo et al. 2012; Barzilay et al. 2013; Gaudart et al. 2013), where almost 10,000 people have died between October 2010 and December 2016. Figure 10.11 reports the initial phase of the epidemic. Precipitation is also shown, because revamping of the disease in Haiti is often associated with heavy rain events.

Early attempts to model WB diseases, in particular cholera dynamics, date back to the work by Capasso and Paveri-Fontana (1979). More recently, Codeço (2001) proposed a simple system of three ordinary differential equations where, in addition to the compartments of susceptible (S) and infected (I) that characterize traditional microparasitological models, one equation accounts for the population dynamics of free-living bacteria (B), i.e. bacteria that can be found in a common water reservoir. Further developments (see, e.g., Rinaldo et al. 2020) have extended the approach to account for space and for other disease characteristics. The standard SIB model can be stated as

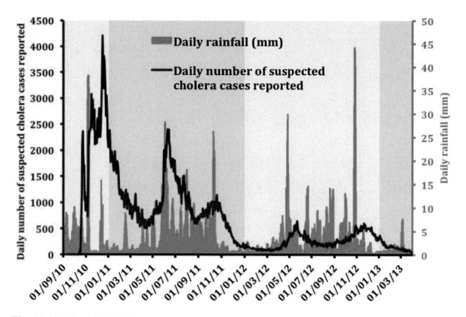

Fig. 10.11 The 2010–2013 course of the cholera epidemic in Haiti plotted together with rainfall events. From Rebaudet et al. (2013)

$$\frac{dS}{dt} = \mu(H - S) - \beta BS$$

$$\frac{dI}{dt} = \beta BS - (\mu + \alpha + \rho)I \qquad (10.11)$$

$$\frac{dB}{dt} = \theta I - \delta B ,$$

where μ is the human natural mortality rate (which is assumed to match the birth rate, so that the human population is at demographic equilibrium in the absence of disease, with H being the population size), β is the rate of exposure to water contaminated by the pathogen, α is the additional mortality rate induced by cholera on infected hosts, ρ is the recovery rate from disease (recovered individuals are assumed to gain long-lasting immunity in this simplified model), θ is the contamination rate (i.e. the rate at which bacteria or viruses shed by infected people reach the water reservoir) and δ is the mortality rate of free-living pathogens. Note that, as a first approximation, we neglect the reproduction of pathogens in the water body. In fact, the only inflow of new pathogens into the water reservoir is the one due to the shedding of infected people.

In addition to the disease-free equilibrium ($S = H, I = 0, B = 0$), system (10.11) has an endemic non-trivial equilibrium with coordinates

$$\bar{S} = \frac{\delta\phi'}{\beta\theta} , \quad \bar{I} = \frac{\mu}{\phi'}\left(H - \frac{\delta\phi'}{\beta\theta}\right) , \quad \bar{B} = \frac{\mu\theta}{\phi'\delta}\left(H - \frac{\delta\phi'}{\beta\theta}\right) ,$$

where $\phi' = \mu + \alpha + \rho$. Of course this equilibrium is biologically meaningful only if $\bar{I}$ and $\bar{B}$ are positive, namely only if

$$H > \frac{\delta(\mu + \alpha + \rho)}{\beta\theta}.$$

By introducing the basic reproduction number

$$R_0 = \frac{\beta\theta H}{\delta(\mu + \alpha + \rho)},$$

it is possible to show that if $R_0 < 1$ the disease-free equilibrium is stable, while the endemic equilibrium is not feasible; conversely, if $R_0 > 1$ the disease-free equilibrium is unstable, while the non-trivial equilibrium is feasible and stable. Note that $1/(\mu + \alpha + \rho)$ is the average residence time of hosts in the infected class, while $1/\delta$ is the average residence time of pathogens in water. Thus, if one infected individual is introduced in a community of size H, he/she will produce $\theta/(\mu + \alpha + \rho)$ pathogens during the infectious period and these pathogens will in turn infect $\beta H/\delta$ susceptibles during the pathogen lifetime in water. Therefore, R_0 can be again interpreted as the average number of secondary infections produced by one infected individual introduced in a healthy population consisting of H individuals. For example, typical reproduction numbers for the ongoing Haitian cholera epidemic range between 2.5 and 4 (Mari et al. 2015), that is they are quite large. Lower values of R_0 have been recorded for other cholera epidemics, e.g. the one that occurred in Zimbabwe during 2008–2009 (Mukandavire et al. 2011).

10.4   Vector-Borne Diseases

These diseases are transmitted by vectors between humans or from animals to humans. The vectors are living organisms, often bloodsucking insects, which ingest disease-producing micro-organisms during a blood meal from an infected host (human or animal) and later inject it into a new host during their subsequent blood meal. Mosquitoes are the best known vectors, but ticks, flies, fleas, snails are also responsible for many of these diseases. Figure 10.12 summarizes their global distribution and nature. Every year there are more than 1 billion cases and over 1 million deaths from vector-borne diseases such as malaria, dengue, schistosomiasis, human African trypanosomiasis, leishmaniasis, Chagas disease, yellow fever, Japanese encephalitis and onchocerciasis, globally. Vector-borne diseases account for over 17% of all infectious diseases (excluding COVID-19).

Malaria is possibly the best-known of these diseases. It is caused by protozoans of the genus *Plasmodium*. It is transmitted by the biting of infected *Anopheles* mosquitoes, which breed in fresh (or less often brackish) water. Because of the importance of water in the development cycle of the parasite vector, malaria is also cata-

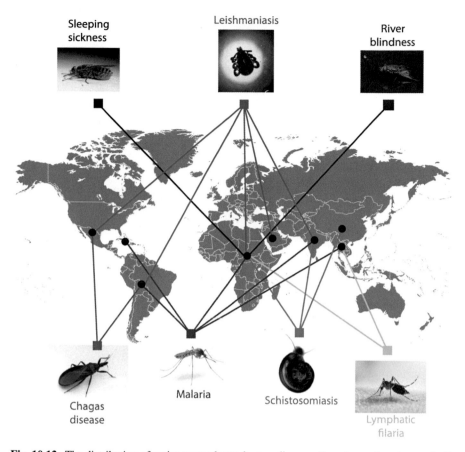

Fig. 10.12 The distribution of major vector-borne human diseases. Parasites and vectors are indicated in Table 10.2, together with the relevant diseases. Redrawn after Matthews (2011). The image of *Biomphalaria glabrata* (schistosomiasis) is taken from Lewis et al. (2008)

logued as a WR disease (World Health Organization 2014b). Malaria occurs mostly in tropical and subtropical countries, particularly in sub-Saharan Africa, SouthEast Asia and South America. Globally, 3.3 billion people are considered to be at risk of infection. According to the latest estimates, there were about 200 million cases in 2013, with more than 580,000 deaths (World Health Organization 2014c). Malaria has been ravaging Italy for centuries. For a long time the disease has been ascribed to "bad air" (*mal aria*, in Italian) until Ronald Ross, a British medical doctor in India, understood that the disease was transmitted by mosquitoes. It was however Giovanni Battista Grassi, an Italian zoologist, who finally identified the *Anopheles* mosquito as the malaria vector (Table 10.2).

Malaria has been the subject of mathematical modelling for more than one century, starting from the pioneering work by Ross (1911). In its original formulation, Ross's (1911) model describes the rates of change of two state variables, i.e. the prevalences

Table 10.2 Parasites and diseases mapped in Fig. 10.12. Modified after Matthews (2011)

Parasite	Disease	Vector	Global distribution
Plasmodium falciparum, Plasmodium vivax, Plasmodium malariae, Plasmodium ovale	Malaria	Anopheline mosquito	Africa, Asia, South America
Trypanosoma cruzi	Chagas disease	Tristomine bugs	South America, Southern USA
Trypanosoma brucei	Sleeping sickness	Tsetse flies	East and west, sub-Saharan Africa
Leishmania major, Leishmania mexica, Leishmania danovani, Leishmania infantum, Leishmania brasilensis	Leishmaniasis	Sandflies	Africa, Asia, South America, Southern USA
Schistosoma haematobium, Schistosoma mansoni, Schistosoma japonicum	Schistosomiasis	Freshwater snails	Africa, Asia, South America
Brugia malayi, Brugia timori, Wurcheraria bancrofti	Lymphatic filariasis	*Anopheles* and *Aedes* mosquitos	Africa, Asia
Onchocerca volvulus	River blindness	Blackflies	Sub-Saharan Africa

of infected humans (U) and infected mosquitoes (M). Demographic dynamics are neglected in both host populations, together with several other features of disease transmission; however, in spite of its simplicity, Ross model still represents the framework of choice for much epidemiological research on malaria (Mandal et al. 2011). In its simplest formulation the model reads as

$$\frac{dU}{dt} = \beta M (1 - U) - \gamma U$$
$$\frac{dM}{dt} = \psi U (1 - M) - \xi M , \tag{10.12}$$

where β is the mosquito-to-human transmission rate, γ is the recovery rate of human hosts, ψ is the human-to-mosquito transmission rate and ξ is the mortality rate of infected mosquitoes. The rationale behind the model is in a way similar to that of Levins (1969) metapopulation model (see Sect. 6.3.2). A fraction 1-U of humans is susceptible to being infected by the protozoans carried by the fraction M of mosquitoes that are infected (like the fraction 1-p of empty patches is susceptible to being colonized by the propagules released by the fraction p of occupied patches). In the same way, the fraction 1-M of susceptible mosquitoes can be infected by the protozoans in the blood of the fraction U of infectious humans. Typical values for γ and ξ are $1.0 \cdot 10^{-2}$ [day^{-1}] and $1.0 \cdot 10^{-1}$ [day^{-1}], that is the average recovery time of infected humans is about 100 days and the average lifetime of infected mosquitoes is about 10 days. Instead β and ψ are quite variable. Actually, β can be considered the product of three factors: $\beta = a \cdot b \cdot c$, where m is the number of female mosquitoes per human host, a is the number of bites per mosquito per unit time, b is the probability of transmission of infection from infectious mosquitoes to humans per bite; ψ as the product ($\psi = ac$) of a and c, the probability of transmission of infection from infectious humans to mosquitoes per bite.

Model (10.12) has two steady states solutions, namely the disease-free equilibrium (DFE), with $U = 0$ and $M = 0$, and the endemic equilibrium (EE), with

$$\bar{U} = \frac{\beta\psi - \gamma\xi}{\psi(\beta + \gamma)}, \qquad \bar{M} = \frac{\beta\psi - \gamma\xi}{\beta(\psi + \xi)}.$$

The EE is feasible if $\beta\psi - \gamma\xi > 0$. Equivalently, the feasibility of the EE is determined by the basic reproduction number

$$R_0 = \frac{\beta\psi}{\gamma\xi}.$$

Specifically, the DFE is stable for $R_0 < 1$, while the EE is feasible and stable for $R_0 > 1$. R_0 can be interpreted as follows. Suppose one infected human is introduced in healthy populations of humans and mosquitoes. Then ψ mosquitoes are infected per unit time. As humans remain infectious for $1/\gamma$ days, this results in ψ/γ mosquitoes being infected. In turn each of these mosquitoes will infect β healthy humans per unit time. As infected mosquitoes stay alive for $1/\xi$ days, each mosquito, before dying, will have infected β/ξ humans. Therefore, one initial infectious human (primary infection) will produce $\frac{\beta\psi}{\gamma\xi}$ secondary infections. Values of R_0 can range from around one to more than 3,000 as estimated by Smith et al. (2007) for 121 African populations.

10.5 Dynamics of Diseases Caused by Macroparasites

In diseases caused by macroparasites, the dynamics of parasites cannot be neglected because their average lifetime is comparable to that of the hosting organisms. Macroparasites grow inside the host but reproduce by releasing infective stages (typically eggs and larvae) into the surrounding external environment, for example through host defecation. The disease transmission occurs then through ingestion by a new host of these infective stages. Many herbivores are for example infected by intestinal worms whose eggs or larvae are ingested during grazing. Here we consider only macroparasites with a simple life cycle, i.e. without intermediate hosts.

An important feature of macroparasites is that their distribution inside the host population can be highly heterogeneous: most of the animals can harbour little or no parasite and a minority can instead accommodate a large number of parasites. Figure 10.13 lists examples of parasite load distributions in various hosts. Since the higher is the parasite load, the greater is the damage inflicted to the host, this heterogeneity must be properly taken into account. It should also be borne in mind that, although there are some exceptions, the death of a host causes the death of the hosted parasites.

The classical model of macroparasite diseases is the one by Anderson and May (1978). Let us denote by H the number or density of the hosts, by P the total number or density of adult parasites inside hosts and by L the number or density of the younger stages of the parasite (e.g. larvae) living outside the hosts. The average parasite load is then P/H. However, the distribution of loads can be heterogeneous and because of this let us denote by p_i the proportion of hosts harbouring i parasites ($i = 0, 1, 2, \ldots$). Note that this implies $\sum_{i=0}^{\infty} i p_i H = P$. Assuming that

(a) new adult parasites are recruited by casual encounters between the free-living infective stages and the hosts,
(b) the mortality of every host linearly increases with the number i of its parasites,
(c) the death of a host causes the death of all adult parasites hosted in it,

one can write the following coupled equations for the host-parasite dynamics:

$$
\begin{cases}
\dot{H} = (\nu - \mu) H - \sum_{i=0}^{\infty} \alpha i p_i H \\
\dot{P} = \beta L H - m P - \sum_{i=0}^{\infty} (\mu + \alpha i) i p_i H
\end{cases}
\tag{10.13}
$$

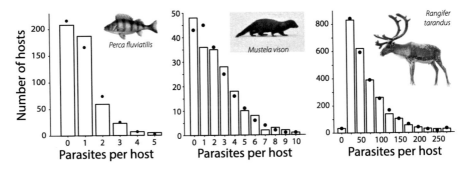

Fig. 10.13 Examples of distributions of the number of macroparasites per host (parasite load) in three different cases: the parasite *Triaenophorus nodulosus* infects internal organs of perch (*Perca fluviatilis*), its intermediate host; parasitic worms of the species *Corynosoma semerne* attach to the intestinal walls of minks (*Mustela vison*); parasitic flies of the species *Hypoderma tarandi* lay their eggs under the skin of the reindeer (*Rangifer tarandus*). Histograms represent data, while points are obtained by suitable fitting of negative binomial distributions. Redrawn after Shaw et al. (1998)

where v and μ are the birth and death rates of hosts in absence of the disease, β is the per unit time probability that a host encounters a larval parasite, m is the mortality rate of the adult parasites inside the host, α is the additional mortality rate due to the presence of each parasite inside the host. The two mortality terms of equations (10.13) that are linked to the parasite load must be properly explained. As for the equation of hosts, it must be noticed that the proportion of hosts containing i parasites (p_i) suffers an additional mortality which amounts to $\alpha \cdot i$ and then the total number of additional deaths per unit time in the host population hosts is in fact $\sum_{i=0}^{\infty} \alpha i p_i H$. As for the equation of parasites, it should be noted that adult parasites do not only die because of the intrinsic mortality m, but also, as we have already specified above, they die together with a dying host that harbours them. Since a host containing i parasites suffers a total mortality $\mu + \alpha i$ and leads to death all the i hosted parasites, it follows that the total number of deaths per unit time in the adult parasite compartment due to this effect equals $\sum_{i=0}^{\infty} (\mu + \alpha i) i p_i H$.

The ultimate form taken by model (10.13) is specified in different ways, depending on the assumptions made for the various demographic and epidemiological parameters that characterize it. As for the microparasitary models, the first important specification concerns the population demography in absence of infection. The two most common assumptions are that hosts follow either a Malthusian or a logistic demography. The second specification relates to the density L of the free-living parasite stages, which live outside the hosts (e.g. in the grass). Obviously, L must be an increasing function of the number of adult parasites that produce these stages (for example as eggs, which are then defecated by hosts) by reproducing inside the host. On the other hand, L must be a decreasing function of the number of hosts, because the greater is H the greater is the ingestion of the parasite free-living stages, with a consequent depressing effect on their density in the surrounding environment. A functional form that is widely used (and that can be justified by a simple model of the L dynamics, which is not reported here) is the following:

$$L = \frac{\theta P}{H + H_0}$$

where θ is the parasite fertility and H_0 is a positive parameter specifying for which values of H the density of the parasite free-living stages is significantly reduced, as a result of ingestion by hosts.

The third and most important specification is that concerning the nature of the distribution p_i of parasite loads inside the host population. A simple statistical distribution that is well suited to a large amount of data, such as those in Fig. 10.13, is the negative binomial distribution (Pielou 1977). It is characterized by two parameters: the mean M and a *clumping* parameter k. Figure 10.14 shows several shapes of the negative binomial distribution for different values of k. The variance σ^2 of the negative binomial distribution is linked to the mean by the following relationship:

$$\sigma^2 = M + \frac{M^2}{k}.$$

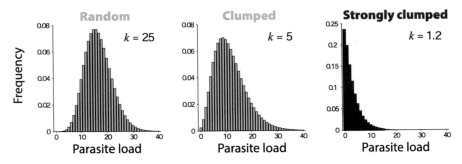

Fig. 10.14 Shape of negative binomial distributions for different values of the clumping parameter k

When the clumping parameter k is large, the variance is almost equal to the mean, as in the Poisson distribution: parasite loads are more or less evenly distributed around the mean. But in most cases k is small (less than 5) and the distribution is clumped, namely a few hosts contain a number of parasites that is much higher than the average parasite load $M = P/H$ and many hosts instead contain quite a small number of parasites. Assuming that the parasite distribution is a negative binomial, we can specify the following terms in equations (10.13)

$$\sum_{i=0}^{\infty} i p_i = M = P/H$$
$$\sum_{i=0}^{\infty} i^2 p_i = M^2 + \sigma^2 = M^2 + M + M^2/k = \frac{P}{H} + \frac{k+1}{k}\frac{P^2}{H^2}$$

(10.14)

Assuming a logistic demography, we can finally formulate the following model for host-macroparasite dynamics:

$$\dot{H} = rH\left(1 - \frac{H}{K}\right) - \alpha P$$
$$\dot{P} = \frac{\lambda P H}{H_0 + H} - (m + \mu + \alpha)P - \alpha\frac{k+1}{k}\frac{P^2}{H}$$

(10.15)

where r and K are, respectively, the intrinsic rate of increase and the carrying capacity of the host population, and $\lambda = \beta\theta$. Like in microparasite models, we now try to determine under which conditions the host population is permanently infested by macroparasites. To this end, we determine the possible equilibria of model (10.15). In addition to the trivial equilibrium ($\bar{H}_0 = 0$, $\bar{P}_0 = 0$), they turn out to be:

- $\bar{H}_K = K$, $\bar{P}_K = 0$;
- $\bar{H}_+ = H^*$ and

$$\bar{P}_+ = \frac{r}{\alpha}H^*\left(1 - \frac{H^*}{K}\right)$$

(10.16)

where H^* is the unique positive solution of the equation

Fig. 10.15 Isoclines and trajectories of the host-macroparasite model (10.15) with logistic demography and influence of the macroparasite on the host survival. Red dashed: $\dot{P} = 0$; Green solid: $\dot{H} = 0$

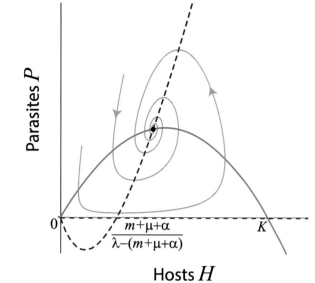

$$\frac{\lambda H}{H_0 + H} - (m + \mu + \alpha) - \frac{k+1}{k} r \left(1 - \frac{H}{K}\right) = 0. \qquad (10.17)$$

It is easy (yet boring) to reduce expression (10.17) to a quadratic equation and determine that there is a positive solution H^* only if $\frac{\lambda}{K+H_0} - (m + \mu + \alpha) > 0$. This inequality can be interpreted in the light of an epidemiological parameters that we have already introduced with reference to microparasite diseases: the basic reproduction number of the disease, R_0. In fact, since the average lifetime of an adult parasite is $1/(m + \mu + \alpha)$, the expected number of secondary infections (that is of new adult parasites) caused by one parasite introduced into a disease-free host population at its carrying capacity turns out to be

$$R_0 = \frac{\lambda K}{H_0 + K} \frac{1}{m + \mu + \alpha}.$$

If $R_0 < 1$ the only non-trivial equilibrium is $(\bar{H}_K, \bar{P}_K)$. It is stable, and so the disease cannot develop. If $R_0 > 1$, the equilibrium $(\bar{H}_+, \bar{P}_+)$ exists and is stable, as shown in Fig. 10.15, which displays the isoclines of model (10.15) and the phase portrait with some trajectories. In this case the disease can develop and it permanently settles into the host population.

It is interesting to analyse how R_0 depends on the different parameters involved: it increases of course with the parasite fertility (parameter λ), it increases and saturates with the host population size K, it decreases with the mortality of both hosts and parasites. In particular, R_0 decreases with the mortality α inflicted by macroparasites to their hosts, and this shows, once again, that very lethal diseases cannot easily settle in animal populations. It is to be remarked that R_0 does not depend on the clumping parameter k. Such a parameter is, however, important in determining the parasite average load: in fact, from Eq. (10.17) we can deduce that H^* decreases with increasing k and from Eq. (10.16) that the average parasite load P^*/H^* decreases with H^*; thus, ultimately, it increases with k. Therefore greater clumping (k small) corresponds to a lower average parasite load.

Until now, we have assumed that the influence of macroparasites on their hosts basically consists of increasing the mortality rate proportionally to the parasite load. However, in many cases, e.g. the already mentioned red grouse population, *Lagopus lagopus scoticus*, hosting the nematode worm *Trichostrongylus tenuis* (Hudson et al. 1992; Dobson and Hudson 1992), the macroparasite exerts its action by decreasing the . We can easily analyse the consequences of this phenomenon by reformulating model (10.13) as follows

$$\dot{H} = \left(\nu - \sum_{i=0}^{\infty} \varepsilon i p_i - \mu \right) H$$
$$\dot{P} = \beta L H - m P - \sum_{i=0}^{\infty} \mu i p_i H \tag{10.18}$$

where we assume that the birth rate decreases proportionally to the parasite load i (with a proportionality constant equal to ε). By making the same assumptions about demography and density of juvenile stages of the parasites that led to formulate model (10.15), we get

$$\dot{H} = r H \left(1 - \frac{H}{K} \right) - \varepsilon P$$
$$\dot{P} = \frac{\lambda P H}{H_0 + H} - (m + \mu) P. \tag{10.19}$$

The analysis of model (10.19) is simple and interesting. In addition to the trivial equilibrium $\bar{X}_0$ and the equilibrium corresponding to the healthy population $\bar{X}_K$, a third equilibrium $\bar{X}_+ = \left[\bar{H} = H^*, \bar{P} = P^* \right]^T$ can exist, at which

$$H^* = \frac{m + \mu}{\lambda - m - \mu} H_0 \qquad P^* = \frac{r}{\varepsilon} H^* \left(1 - \frac{H^*}{K} \right).$$

It is easy to see that this equilibrium is feasible ($H^* > 0$, $P^* > 0$) if the condition holds that the basic reproduction number of the disease R_0 is greater than unity, where

$$R_0 = \frac{\lambda K}{H_0 + K} \frac{1}{m + \mu}.$$

Even if feasible, however, this equilibrium is not always stable. In particular, it can be shown that it is unstable if $H^* < K/2$. The two situations ($H^* > K/2$ and $H^* < K/2$) are illustrated in Fig. 10.16 through the graph of isoclines and a sketch of some trajectories. In the case in which $H^* < K/2$ there is no stable equilibrium and the host and parasite populations converge toward a stable limit cycle (periodic solution) of appropriate period. Note that this situation can establish when the parasite

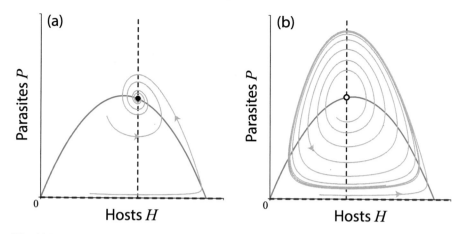

Fig. 10.16 Isoclines and trajectories of model (10.19) in the case of **a** stable steady state and **b** periodic regime with a limit cycle. Red dashed: $\dot{P} = 0$; Green solid: $\dot{H} = 0$

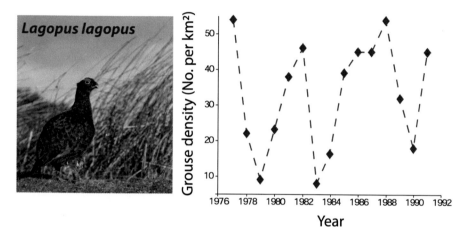

Fig. 10.17 Results of the study by Hudson et al. (1998) concerning the influence of macroparasites on the red grouse *Lagopus lagopus scoticus* demography: evolution of the numbers of grouse (shot and counted) in the north of England

fertility θ or the contact rate β are large (which implies a large λ) or when the half-saturation constant H_0 is small. Of course, if $R_0 < 1$ the equilibrium $\bar{X}_+$ is not feasible, and the disease cannot establish in the population so that the only stable equilibrium is the one corresponding to the disease-free population $\bar{X}_K$.

Possible self-sustained oscillations obtained with model (10.19) are corroborated by evidence from the red grouse population study. The statistics from this bird hunting bags show that the grouse has a demographic cycle of 4–7 years since the 19th century (see Fig. 10.17). Hudson et al. (1998) experimentally showed that these fluctuations are due to the parasite *Trichostrongylus tenuis*: in English moors where grouse have

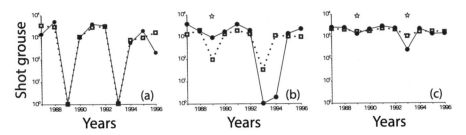

Fig. 10.18 Numbers of shot grouse in two populations of northern England that were either treated or untreated with anthelminthic drugs. **a** untreated, **b** treated once, and **c** treated twice; asterisks indicate years in which anthelminthics have been used. Redrawn after Hudson et al. (1998)

been treated with anthelminthic medication for more than a decade, demographic fluctuations have indeed disappeared, as shown in Fig. 10.18.

10.6 Host-Parasitoid Dynamics

As we stated in the introduction, parasitoids play an important role in applied ecology, because they are often used as an effective substitute for pesticides. This practice is termed biological control. Many harmful organisms, in fact, turn out to be potential hosts of parasitoids, which are then released into the environment to clear crops from pest. The previous descriptions of parasite dynamics have been performed by assuming that reproduction occurs continuously over time and then by means of differential equations. In the case of host-parasitoid interactions, however, we must necessarily resort to discrete-time models. In fact, both organisms are usually insects and are therefore characterized by concentrated breeding and non-overlapping generations.

The simplest approach to host-parasitoid dynamics is due to Nicholson (1933) and to Nicholson and Bailey (1935). Let us define

H_k = number of hosts in generation k
P_k = number of adult parasitoids in generation k
A_k = numbers of attacked hosts in generation k

Assuming that, in absence of parasitoids, hosts have a Malthusian dynamics with finite rate of increase λ and that adult parasitoids are a constant fraction σ of the young parasitoids emerging from the hosts, one gets

$$H_{k+1} = \lambda(H_k - A_k)$$
$$P_{k+1} = \sigma A_k$$

(10.20)

Note that the number of emerging parasitoids equals the number of attacked hosts, because—although a host may encounter more than one parasitoid—only the egg laid during the first encounter will mature at the expense of the host. This observation allows calculation of A_k. In fact, let T be the time interval during which parasitoids are active; then the number of attacked hosts is

$$A_k = H_k p(T)$$

where $p(t)$ indicates the probability that a host makes one or more encounters over the time period t. This probability can easily be evaluated by using the classical hypothesis of Poisson-distributed events. Under the assumption that the probability of making two or more encounters during an infinitesimal time interval dt is negligible, while the probability of encountering exactly one parasitoid during that infinitesimal time is proportional to dt and to the number of parasitoids, we obtain

$$p_0(t + dt) = p_0(t)(1 - \alpha P_k dt)$$

where α is a constant of proportionality and $p_0(t)$ is the probability of making no encounters during time t. We then get

$$\dot{p}_0 = -\alpha P_k p_0$$

and, by integrating and assuming the obvious condition $p_0(0) = 1$,

$$p_0(t) = \exp(-\alpha P_k t).$$

Therefore, the probability of making one or more encounters during the time interval T is given by

$$p(T) = 1 - p_0(T) = 1 - \exp(-\alpha T P_k)$$

from which, by setting $\beta = \alpha T$, we get the expression for the number of attacked hosts:

$$A_k = H_k \left(1 - e^{-\beta P_k}\right).$$

Substituting into Eq. (10.20) we finally obtain

$$\begin{aligned} H_{k+1} &= \lambda H_k \exp\left(-\beta P_k\right) \\ P_{k+1} &= \sigma H_k \left(1 - \exp(-\beta P_k)\right). \end{aligned} \tag{10.21}$$

These equations by Nicholson and Bailey constitute the basic model for studying the dynamics of host and parasitoid populations. The equilibria are obtained by imposing the usual conditions

$$\begin{aligned} H_{k+1} &= H_k = H \\ P_{k+1} &= P_k = P. \end{aligned} \tag{10.22}$$

They are therefore solutions of the equations

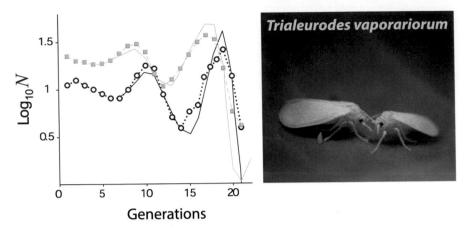

Fig. 10.19 Fluctuations of a population of *Trialeurodes vaporariorum* (filled squares) and its parasitoid *Encarsia formosa* (empty circles), and results (thin lines) of simulations obtained by fitting Nicholson and Bailey's model to data. Redrawn after Burnett (1958)

$$H = \lambda H \exp(-\beta P)$$
$$P = \sigma H (1 - \exp(-\beta P)) \tag{10.23}$$

and are given by

$$\bar{H}_0 = 0, \qquad \bar{P}_0 = 0 \tag{10.24}$$

and

$$\bar{H}_+ = \frac{\lambda}{\sigma \beta} \frac{\ln(\lambda)}{\lambda - 1} \qquad \bar{P}_+ = \frac{1}{\beta} \ln(\lambda) \tag{10.25}$$

Equilibrium (10.24) is trivial and is obviously unstable, while at equilibrium (10.25) we may have coexistence of hosts and parasitoids. However, it can be proved that this non-trivial equilibrium is also unstable. The model by Nicholson and Bailey, unlike that of Lotka and Volterra for predator-prey dynamics, of which it is the discrete time counterpart, predicts fluctuations in the two populations that are gradually becoming larger and larger. This is obviously due to the Malthusian assumption for the host demography. The result shows anyway that, unlike predators and microparasites, parasitoids cannot regulate a Malthusian population. If Nicholson and Bailey's model seems to describe well enough some real situations for a limited number of generations (as shown in Fig. 10.19), it is not acceptable for describing long-term dynamics. Figure 10.2 shows, for example, that *Heterospilus prosopidis* and *Callosobruchus chinensis* exhibit permanent periodic oscillations.

More realistic models than Nicholson and Bailey's can explain different behaviours of host-parasitoid systems and include phenomena such as intraspecific competition of hosts and the functional response of parasitoids. We do not further elaborate on this topic: the interested reader can refer to the monograph by Hassell (1978).

References

Anderson RM, May RM (1978) Regulation and stability of host-parasite population interactions: I. Regulatory processes. *Journal of Animal Ecology* 47:219–247

Anderson RM, May RM (1979) Population biology of infectious diseases: Part I. *Nature* 280:361–367

Anderson RM, May RM (1991) *Infectious Diseases of Humans: Dynamics and Control.* Oxford University Press, Oxford, UK

Barzilay EJ, Schaad N, Magloire R, Mung KS, Boncy J, Dahourou GA, Mintz ED, Steenland MW, Vertefeuille JF, Tappero JW (2013) Cholera surveillance during the Haiti epidemic—The first 2 years. *New England Journal of Medicine* 368:599–609

Bernoulli D (1760) Essai d'une nouvelle analyse de la mortalité causé par la petite vérole et des advantages de l'inoculation pour la preévenir. *Academie Royale des Sciences: Histoire et Memoires de Mathematique et de Physique*, pp 1–45

Burnett T (1958) Dispersal of an insect parasite over a small plot. *Canad Entomol* 90:279–283

Capasso V, Paveri-Fontana S (1979) A mathematical model for the 1973 cholera epidemic in the European Mediterranean region. *Revue Epidemiologique de Santé Publique* 27:121–132

Codeço C (2001) Endemic and epidemic dynamics of cholera: The role of the aquatic reservoir. *BMC Infectious Diseases* 1:1

Daszak P, Cunningham AA, Hyatt AD (2000) Emerging infectious diseases of wildlife—Threats to biodiversity and human health. *Science* 287:443–449

Dobson A (1988) Behavioural and life history adaptations of parasites for living in desert environments. *Journal of Arid Environment* 17:185–192

Dobson A, Hudson PJ (1992) Regulation and stability of a free-living host-parasite system, *Trichostrongylus tenuis* in red grouse. II: Population models. *Journal of Animal Ecology* 61:487–498

Dobson A, MacCallum H (1997) The role of parasites in bird conservation. In: Clayton DH, Moore J (eds) *Host-Parasite Evolution: General Principles and Avian Models.* Oxford University Press, Oxford, UK, pp 155–173

Gaudart J, Rebaudet S, Barrais R, Boncy J, Faucher B, Piarroux M, Magloire R, Thimothc G, Piarroux R (2013) Spatio-temporal dynamics of cholera during the first year of the epidemic in Haiti. *PLOS Neglected Tropical Diseases* 7:1–10

Hassell M (1978) *The Dynamics of Arthropod Predator-Prey Systems.* Princeton University Press, Princeton

Heesterbeek J, Roberts M (1995) Mathematical models for microparasites of wildlife. In: Grenfell B, Dobson A (eds) *Ecology of Infectious Diseases in Natural Populations.* Cambridge University Press, Cambridge, UK, pp 90–122

Hudson P, Dobson A, Lafferty KD (2006) Is a healthy ecosystem one that is rich in parasites? *Trends in Ecology and Evolution* 21:381–386

Hudson P, Dobson A, Newborn D (1992) Regulation and stability of a free-living host-parasite system: *Trichostrongylus tenuis* in red grouse. I. Monitoring and parasite reduction experiments. *Journal of Animal Ecology* 61:477–486

Hudson P, Dobson A, Newborn D (1998) Prevention of population cycles by parasite removal. *Science* 282:2256–2258

Iannelli M, Pugliese A (2014) *An Introduction to Mathematical Population Dynamics.* Springer, Switzerland

Jones KE, Patel NG, Levy MA, Storeygard A, Balk D, Gittleman JL, Daszak P (2008) Global trends in emerging infectious diseases. *Nature* 451:990–994

Levins R (1969) Some demographic and genetic consequences of environmental heterogeneity for biological control. *Bulletin of the Entomogical Society of America* 15:237–240

Lewis FA, Liang Ys, Raghavan N, Knight M (2008) The NIH-NIAID schistosomiasis resource center. *PLOS Neglected Tropical Diseases* 2:1–4

Lipkin W (2013) The changing face of pathogen discovery and surveillance. *Nature Reviews Microbiology* 11:133–141

Mandal S, Sarkar RR, Sinha S (2011) Mathematical models of malaria—A review. *Malaria Journal* 10:202

Mari L, Bertuzzo E, Finger F, Casagrandi R, Gatto M, Rinaldo A (2015) On the predictive ability of mechanistic models for the Haitian cholera epidemic. *Journal of the Royal Society Interface* 12:20140840

Matthews KR (2011) Controlling and coordinating development in vector-transmitted parasites. *Science* 331:1149–1153

May R, Anderson R (1979) Population biology of infectious diseases: Part II. *Nature* 280:455–461

Morens D, Fauci AS (2020) Emerging pandemic diseases: How we got to COVID-19. *Cell* 182:1077–1092

Morens D, Folkers G, AS F, (2004) The challenge of emerging and re-emerging infectious diseases. *Nature* 430:242–249

Mukandavire Z, Liao S, Wang J, Gaff H, Smith DL, Morris JG (2011) Estimating the reproductive numbers for the 2008–2009 cholera outbreaks in Zimbabwe. *Proceedings of the National Academy of Sciences* 108:8767–8772

Nicholson A (1933) The balance of animal populations. *Journal of Animal Ecology* 2:131–178

Nicholson A, Bailey V (1935) The balance of animal populations. Part I. *Proc Zool Soc Lond* 551–598

Pacini F (1854) Osservazioni microscopiche e deduzioni patologiche sul cholera asiatico. *Gazzetta medica italiana—Toscana* 4(397–401):405–12

Pielou E (1977) *Mathematical Ecology*. Wiley, New York, NY

Plowright W (1985) La peste bovine aujourd'hui dans le monde. Controle et possibilité d'eradication par la vaccination. *Annales de Médecine Vétérinaire* 129:9–32

Rebaudet S, Gazin P, Barrais R, Moore S, Rossignol E, Barthelemy N, Gaudart J, Boncy J, Magloire R, Piarroux R (2013) The dry season in Haiti: A window of opportunity to eliminate cholera. *PLOS Currents Outbreaks* Jun 10. Edition 1

Rinaldo A, Bertuzzo E, Mari L, Righetto L, Blokesch M, Gatto M, Casagrandi R, Murray M, Vesenbeckh S, Rodriguez-Iturbe I (2012) Reassessment of the 2010–2011 Haiti cholera outbreak and rainfall-driven multiseason projections. *Proceedings of the National Academy of Sciences USA* 106:6602–6607

Rinaldo A, Gatto M, Rodriguez-Iturbe I (2020) *River Networks as Ecological Corridors. Species, Populations, Pathogens*. Cambridge University Press, New York

Roepstorff A, Nansen P (1998) Epidemiology, diagnosis and control of helminth parasites of swine. Technical report. FAO, Rome

Ross R (1911) *The Prevention of Malaria*. John Murray, London, UK

Scott G (1964) Rinderpest. *Advances in Veterinary Science,* 9:113–224

Scott M, Dobson A (1989) The role of parasites in regulating host abundance. *Parasitology Today* 5:176–183

Shaw DJ, Grenfell BT, Dobson A, (1998) Patterns of macroparasite aggregation in wildlife host populations. *Parasitology* 117:597–610

Siles-Lucas M, Becerro-Recio D, Serrat J, González-Miguel J (2021) Fascioliasis and fasciolopsiasis: Current knowledge and future trends. *Research in Veterinary Science* 134:27–35

Smith DL, McKenzie FE, Snow RW, Hay SI (2007) Revisiting the basic reproductive number for malaria and its implications for malaria control. *PLoS Biology* 5:e42

Utida S (1957) Cyclic fluctuations of population density intrinsic to the host parasite system. *Ecology* 38:442–449

Waldor MK, Ray Chaudhuri D (2000) Bacterial genomics: Treasure trove for cholera research. *Nature* 406:469–470

World Health Organization (2014a) Global health observatory. Technical report. World Health Organization, Geneva, Switzerland

World Health Organization (2014b) Preventing diarrhoea through better water, sanitation and hygiene. Technical report. World Health Organization, Geneva, Switzerland

World Health Organization (2014c) World malaria report 2014. Technical report. World Health Organization, Geneva, Switzerland

Chapter 11
Problems on the Ecology of Parasites and Disease

Problem PD1

Rabies is a serious microparasitic disease due to the transmission of a virus between hosts. One of the most important animal hosts is the red fox (*Vulpes vulpes*). Write down an SI model for this mammal assuming that demography is logistic, transmission is density-dependent and no infected animal can recover from the disease.

Use the following parameters:

- birth rate $v = 0.6$ year^{-1};
- natural death rate $\mu = 0.2$ year^{-1};
- carrying capacity $K = 5$ foxes km^{-2};
- disease-related death rate $\alpha = 5$ year^{-1};
- basic reproduction number of rabies $R_0 = 3$.

Compute:

(a) the transmission coefficient β of rabies in foxes;
(b) the endemic equilibrium densities of susceptible and infected foxes;
(c) the equilibrium prevalence of the disease.

Problem PD2

The *gonorrhoea* of the fantasy population of striped kangaroos is a bacterial disease that strikes these kangaroos as a consequence of sexual contacts. The contact rate increases with the number of kangaroos and then saturates, because the maximum number of sexual contacts per unit time is obviously finite. The disease does not confer immunity. Describe the gonorrhoea dynamics in Marsupiumland by an SI model with saturating transmission and characterized by

- birth rate $= 0.2$ year^{-1};

© The Author(s), under exclusive license to Springer Nature Switzerland AG 2022
M. Gatto and R. Casagrandi, *Ecosystem Conservation and Management*,
https://doi.org/10.1007/978-3-031-09480-4_11

Problem PD1   The red fox

- natural death rate = $0.05 \, \text{year}^{-1}$;
- carrying capacity = 50 kangaroos km^{-2};
- disease-related death rate = $0.01 \, \text{year}^{-1}$;
- infection rate = $\beta I / (\delta + N)$, with $\beta = 3 \, \text{year}^{-1}$ and $\delta = 10$ kangaroos km^{-2} (with $N = S + I$);
- recovery rate = $2 \, \text{year}^{-1}$.

Based on these data:

(i)   write down the model for the gonorrhoea dynamics;
(ii)   compute the basic reproduction number of the disease;
(iii)   determine whether gonorrhoea can permanently establish within Marsupium-land kangaroos; and
(iv)   compute the prevalence at the endemic equilibrium.

Problem PD3

In many microparasitic diseases a fraction of infected and infectious individuals is not symptomatic and thus can reproduce. For instance, *cholera* and *amoebiasis* are diseases with this peculiarity. Analyse the effect of asymptomatic individuals in an SI system without immunity, with density-dependent transmission and Malthusian demographics. Assume that:

- the birth rate of susceptibles and asymptomatics is $\nu = 0.7 \, \text{year}^{-1}$;
- the natural death rate is $\mu = 0.2 \, \text{year}^{-1}$;
- the disease-related death rate is $\alpha = 0.01 \, \text{year}^{-1}$;

- S and I are measured as No. of individuals km^{-2};
- the transmission coefficient from infected to susceptible is $\beta = 2\,\text{year}^{-1}$ (No. of individuals km^{-2})$^{-1}$;
- the average recovery time from the disease is 15 days;
- a fraction σ of infected is symptomatic and cannot reproduce.

Write down the SI model equations and determine how the model equilibria vary for increasing σ. Find out the values of σ for which the disease can demographically regulate the Malthusian population.

Problem PD4

Selective culling is a method that can be employed to try to control wildlife diseases (e.g. *rabies*). Perform a simple analysis of the efficacy of this control method by using an SI model without immune response and with density-dependent transmission. Assume the following values for the parameters

- birth rate of susceptibles $v = 1.5\,\text{year}^{-1}$;
- natural death rate $\mu = 0.5\,\text{year}^{-1}$;
- disease-related death rate $\alpha = 1\,\text{year}^{-1}$;
- S and I are measured as No. of individuals km^{-2};
- carrying capacity $K = 13$ individuals km^{-2};
- transmission coefficient from infected to susceptible $\beta = 1\,\text{year}^{-1}$ (No. of individuals km^{-2})$^{-1}$;
- average recovery time from the disease = 2 months.

Find out whether the disease can establish in the population. If it can, analyse whether culling can eradicate the disease. Assume that hunters can distinguish and kill the infected animals only, inflicting a death rate h (year^{-1}). How big should h be to permanently eliminate the disease from the wildlife population?

Problem PD5

Classical swine fever (CSF) is a viral disease with severe economic consequences for wild boars and domestic pigs. Domestic pigs as well as wild boars are highly susceptible to CSF infection. There are well documented reports that CSF may spill over from wild boar to domestic pigs. Transmission of the infection is by direct contact and the transmission is density-dependent.

Analyze the dynamics of *classical swine fever* in wild boars (*Sus scrofa*) using the information provided by Hone et al. (1992) who studied the disease in a Pakistan boar population. The demography is logistic with

- carrying capacity $K = 10.4$ ind. km^{-2};

Problem PD5 A wild boar
in Pakistan

Sus scrofa

- instantaneous intrinsic rate of demographic increase $r = 0.09$ year^{-1};
- natural mortality rate $\mu = 0.6$ year^{-1}.

The fever is very virulent with a quite high mortality rate due to the disease, namely $\alpha = 0.2$ day^{-1}. The recovery period for the few animals that survive is 15 days and the immunity is permanent. The estimated coefficient of disease transmission is $\beta = 0.044$ (ind. km^{-2})$^{-1}$ day^{-1}. Assume that the number of recovered boars can be approximately set to zero because mortality due to disease is very high.

Then

(a) Write a simple SI model for the disease;
(b) Calculate the basic reproduction number of CSF.

One of the possible methods for curbing the disease is by culling. Assume that both susceptible and infected boars are killed at a rate c (year^{-1}). Calculate then the value of c above which the disease cannot become endemic.

Problem PD6

Vaccination is the most important means for preventing disease epidemics. Analyse its efficacy by considering a simple case of a city whose demography is described by a constant flow w of births and immigration and a mortality rate μ. Consider a microparasitic disease with density-dependent transmission that provides full immunity to people that recover. The population parameters are

- $w = 15000$ people year^{-1};

- μ = death rate of susceptible people = 0.015 year^{-1};
- α = additional mortality rate due to the disease = 0.005 year^{-1};
- β = coefficient of disease transmission = 0.0001 (number of people)$^{-1}$ year^{-1};
- recovery time = 0.5 months.

Assume that susceptible individuals are vaccinated at a rate V, where V is expressed as year^{-1}. Then

(a) Write down the SI model that governs the epidemiological dynamics;
(b) Calculate the basic reproduction number with no vaccination;
(c) Calculate how big the vaccination rate V should be in order to avoid that the disease establishes in the population;
(d) Calculate the number of susceptible people at equilibrium in the case of successful vaccination.

Problem PD7

Cholera is a microparasitic disease transmitted through exposure to water contaminated by the bacterium *Vibrio cholerae*. In the standard SIB model the rate at which one susceptible becomes infected is assumed to linearly increase with the bacterial concentration. A more realistic assumption is that it increases and saturates with the concentration in the following way:

$$\text{infection rate } = \beta B/(1 + B)$$

where B is the normalized concentration of bacteria in the water (namely the concentration of bacteria divided by their carrying capacity). It is thus possible to better analyze the dynamics of *cholera* in a human community in which the recruitment of new susceptible individuals (due to either newborns or migration from nearby communities) is a constant flow w and the susceptible individuals have a mortality rate μ. For the modified SIB model assume that recovered people are permanently immune and parameters have the following values

- average life time of susceptibles: 70 years;
- constant flow $w = 5$ individuals day^{-1};
- infection rate = $\beta B/(1 + B)$ with $\beta = 1$;
- recovery rate $\rho = 0.2$ day^{-1};
- contamination rate $\theta = 10^{-6}$ (No.infected)$^{-1}$ day^{-1};
- death rate of *V. cholerae* $\delta = 0.2$ day^{-1}.

Based on these data, write down the modified SIB model for the cholera dynamics.
Then, consider a population that at the endemic equilibrium is characterized by 10,000 infected individuals. For this population

(a) Compute the equilibrium prevalence of the disease;
(b) Determine the disease-related death rate α.

Problem PD8

Zika virus is an emerging mosquito-borne virus that infects and causes disease in humans. Although symptoms are usually mild, there is evidence that the infection of women during a critical part of pregnancy can lead to the development of micro-cephaly in the unborn child. The most important disease vector is the Yellow Fever mosquito, *Aedes aegypti*. Caminade et al. (2017) have studied the dynamics of the disease in relation to the climate of the countries hit by an epidemic of the virus. They estimated the following parameters for a country with an average temperature of 25 °C.

- a = number of bites per mosquito per unit time = $0.2 \, \text{day}^{-1}$
- m = number of female mosquitoes per human host = 50
- b = probability of transmission of infection from infectious mosquitoes to humans per bite = 0.5
- ξ = mortality rates of mosquitoes = $0.19 \, \text{day}^{-1}$
- recovery time from the disease = 7 days
- c = probability of transmission of infection from infectious humans to mosquitoes per bite = 0.1

You are required to:

(a) From the above parameters derive β, the mosquito-to-human transmission rate and ψ, the human-to-mosquito transmission rate.
(b) Write down a Ross model for the Zika virus describing the dynamics of the prevalence of infected humans (U) and that of infected mosquitoes (M).

Problem PD8   The Yellow Fever mosquito, the most important disease vector for the Zika virus

Aedes aegypti

(c) Calculate the basic reproduction number and assess whether the disease can establish in the country. If it can, calculate the prevalence of both humans and mosquitoes at equilibrium.

Problem PD9

West Nile virus (WNV) is a mosquito-borne flavivirus which has caused repeated outbreaks in humans in southern and central Europe. The main vector for WNV is the common mosquito *Culex pipiens*. Although the disease can be transmitted to humans the main virus reservoir is represented by birds. Vogels et al. (2017) have studied how the basic reproduction number of the disease varies with the average temperature of the country hosting mosquitoes and birds. They estimated the following parameters for countries with an average temperature of 23 °C and 28 °C, respectively.

- a = number of bites per mosquito per unit time = $0.14\,\text{day}^{-1}$ at 23 °C and $0.2\,\text{day}^{-1}$ at 28 °C
- m = number of female mosquitoes per bird host = 10
- b = probability of transmission of infection from infectious mosquitoes to birds per bite = 0.8
- average lifetime of mosquitoes = 33 days at 23 °C and 25 days at 28 °C
- bird recovery time from the disease = 5.5 days
- c = probability of transmission of infection from infectious birds to mosquitoes per bite = 0.04 at 23 °C and 0.34 at 28 °C

(A) From the above parameters derive β, the mosquito-to-bird transmission rate and ψ, the bird-to-mosquito transmission rate for both temperatures.
(B) Write down a Ross model for WNV describing the dynamics of the prevalence of infected birds (U) and that of infected mosquitoes (M).

Problem PD9 The common mosquito, the most important vector for West Nile Virus

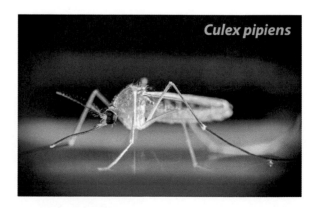

Culex pipiens

(C) Calculate the basic reproduction number and establish whether the disease can establish at the two temperatures. If it can, calculate the prevalence of both birds and mosquitoes at equilibrium.

Problem PD10

Lyme disease, also known as Lyme borreliosis, is an infectious vector-borne disease caused by the *Borrelia spp.* bacterium. It is transmitted to humans and to many other mammals by the bites of infected ticks of the genus *Ixodes spp.* that leave a typical rash on the skin of bitten people, as shown in the Figure.

However, humans are not the main hosts of Lyme disease. A growing body of evidence implicates small mammals (e.g. mice, chipmunks, shrews) as key hosts of the disease. Moreover, a lot of other animals (e.g. deer) can be bitten by ticks but are not able to infect them, namely, even if they were bitten by infected ticks, they would not infect healthy ticks with borreliosis. This kind of hosts are called incompetent hosts and the phenomenon is termed dilution. In fact, if there are many incompetent hosts around, many of the ticks will make their blood meal on them and will not transmit the disease.

Study the effect of incompetent hosts by assuming the following realistic data for the small mammals (host) and ticks (vector) system:

- m_0 = mean number of ticks per small mammal when there are no incompetent hosts around = 0.15

Problem PD10 The rash due to Lyme disease on the leg of a person bitten by a tick of the species *Ixodes ricinus* (inset)

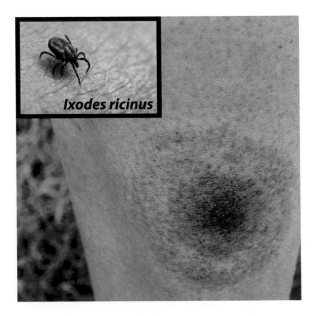

Ixodes ricinus

- a = mean number of bites per tick per day = $0.4\,\text{day}^{-1}$
- b = probability of transmission from tick to small mammal = 0.9
- c = probability of transmission from small mammal to tick = 0.8
- ξ = tick mortality = $0.015\,\text{day}^{-1}$
- γ = recovery rate of small mammals = $0.3\,\text{day}^{-1}$

When there are D [No. individuals km^{-2}] incompetent hosts in the environment, then the mean number of ticks per small mammal is lower and given by

$$m = \frac{m_0}{1 + \epsilon D}$$

with $\epsilon = 0.05$ (No. individuals km^{-2})$^{-1}$.

(i) calculate the basic reproduction number of the disease without incompetent hosts;
(ii) calculate the prevalence of ticks and hosts at equilibrium;
(iii) calculate the density D of incompetent hosts above which Lyme disease cannot establish in the small mammals.

Problem PD11

Fish populations can harbour several species of macroparasites. For instance the yellow perch (*Perca fluviatilis*) can be infected by the cestode worm *Triaenophorus nodulosus*. The characteristics of the parasite load distribution in the perch is shown in the Figure which reports the mean parasite load and the clumping parameter k. The yellow perch lives 5 years in the average. The mortality inflicted to the fish by the macroparasite is not easy to quantify, but one can approximately assume that a parasite load of 5 worms per fish inflicts a mortality which is about half the natural mortality of the perch. The estimated carrying capacity of the perch (e.g. in Lake Varese, northern Italy) is about 6,000 individuals km^{-2}. Its intrinsic instantaneous rate of increase is approximately $0.05\,\text{year}^{-1}$. The average life time of the adult *T. nodulosus* is not exactly known but can be assumed to be about 2 years.

Assume that the basic reproduction number of the disease is 1.2 and that the average parasite load reported in the Figure is the one that would establish at the equilibrium between hosts and parasites. By using the model by Anderson and May (1978), evaluate the abundance of hosts at equilibrium. Also, estimate the values of the two parameters λ and H_0 in the relationship $\lambda H/(H_0 + H)$ that links the fertility of one adult worm to the host density.

Problem PD11 Distribution
of *Triaenophorus nodulosus*
loads inside the yellow perch
Perca fluviatilis

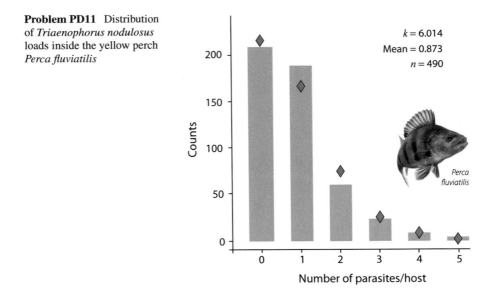

Problem PD12

The blue partridge of the Po valley (*Lagopus padanus*, a fantasy species) is often infested by nematode worms of the genus *Trichonicus*, which are not lethal, but have noxious effects on the bird fertility. When the partridge population is disease-free, the carrying capacity is $K = 40$ individuals km^{-2}, while the death rate is $\mu = 0.2$ year^{-1} and the intrinsic rate of demographic increase is $r = 0.1$ year^{-1}. Let L be the parasite load (i.e. the average number of adult worms inside the guts of each partridge); then the decrease of the per capita natality rate is εL with $\varepsilon = 0.05$ (No. of parasites/No. of partridges)$^{-1}$ year^{-1}. The death rate m of the parasite is 0.6 year^{-1}, while its reproduction rate (year^{-1}) is a function of the density H of partridges: $2H/(10 + H)$.

Find the coexistence equilibrium between parasites and hosts. Then graph the isoclines and determine the epidemiological dynamics.

References

Anderson RM, May RM (1978) Regulation and stability of host-parasite population interactions: I. Regulatory processes. *Journal of Animal Ecology* 47:219–247

Caminade C, Turner J, Metelmann S, Hesson JC, Blagrove MSC, Solomon T, Morse AP, Baylis M (2017) Global risk model for vector-borne transmission of zika virus reveals the role of El Niño 2015. *Proceedings of the National Academy of Sciences* 114:119–124

Hone J, Pech R, Yip P (1992) Estimation of the dynamics and rate of transmission of classical swine fever (hog cholera) in wild pigs. *Epidemiology and Infection* 108:377–386

Vogels C, Hartemink N, Koenraadt C (2017) Modelling west nile virus transmission risk in Europe: Effect of temperature and mosquito biotypes on the basic reproduction number. *Scientific Reports* 7:5022

Index

© The Editor(s) (if applicable) and The Author(s), under exclusive license
to Springer Nature Switzerland AG 2022
M. Gatto and R. Casagrandi, *Ecosystem Conservation and Management*,
https://doi.org/10.1007/978-3-031-09480-4

Species Index

Author Index

Printed in the United States
by Baker & Taylor Publisher Services